Laser

F. J. Duarte

To Jeremy

October 16, 2012.

Rochester.

First published in 2012

OPTICS JOURNAL
P. O. Box 26592
Rochester
New York 14626
U S A

http://www.opticsjournal.com

International Standard Book Number (Print): 978-0-9760383-1-3

International Standard Book Number (Electronic): 978-0-9760383-2-0

Library of Congress Control Number: 2012904907

Other titles by Optics Journal:

Ward, J. C. (2004). *Memoirs of a Theoretical Physicist*, Optics Journal.

In memory of my father… *Luis Enrique* (1921-2011).

Contents

Chapter 8: Reflections I

Chapter 9: Reflections II

Preface

Laser Physicist is mainly an exposition of facts and opinions from my career in laser physics, that begun at Macquarie University, Sydney Australia. Besides some background, the story begins with my 1971 arrival, as a teenage boy, to Sydney and continues with recollections from my progress through the academic life at Macquarie University. The focus then shifts to laser research in America, first in academia, and then in military, and industrial research. As such, a first hand account of the dynamics driving laser research during the Cold War years is provided. Those were, in my opinion, the golden years of laser development. Recollections of my interaction with some giants of the quantum era, are also included.

The contrast between academic and industrial research is highlighted with an insider's view of the glory days at an American icon. This is a detailed account of the internal dynamics of R&D within the framework of an industrial colossus, at its pinnacle, followed by a description of its decline, and the various factors that might have led to its downfall, and present prostration.

After 33 years as a practicing laser physicist, and having experienced first hand, *the gift of discovery,* I venture into a series of briefs essays on various subjects of profound interest to the curious mind. Discussions on quantum mechanics, reality, the need to explore, divinity, and the emergence of nanogods, are included. In this regard, the opinions on divinity of Dirac, Feynman, Heisenberg, Maxwell, Newton, Schrödinger, and other luminaries, are examined and discussed from a personal perspective using clear and transparent language. A foreword is not included since those deemed qualified to write such piece, are not here anymore.

F. J. Duarte
Rochester, New York.
April, 2012

Chapter 1

Santiago

1. Santiago and travels

1. Santiago and travels

I was born in Santiago de Chile, a city founded in 1540 by the Spanish soldier and conquistador don Pedro de Valdivia.

My parents are Luis and Ruth. My brother's name is Henry. Our forefathers arrived in Chile, from Spain, in the period 1540-1600. Family life was very strong and important, in Chilean society in the 1960s, when I was growing up. Not only the family as a nucleus was important but there was also a very active interaction with immediate family members, uncles, cousins, and the like. I still remember very fondly many sunny Sundays when visitors would come to the house of my maternal grand parents (Francisco and Rosa) and enjoy long conversations drinking tea after a traditional Spanish-style lunch. My early childhood took place in Santiago. I have many happy memories growing up and traveling extensively along Chile.

Since my days in kindergarten I tended to show a rebellious attitude towards rituals and authority. Consequently, to the dismay of my parents, in my early schooling I was expelled from several schools. In my second attempt at kindergarten, one day that the teacher was absent, I organized the escape of the whole class (in an all-boys school) through a window while piling up the furniture against the doors of the classroom to prevent the school director (nicknamed "gorilla") from coming in. At the end of the school year my father went to pick up the school report that was all written in red with the lowest possible marks. I still remember clearly the teacher talking and talking to my father whose face at time would turn red. Once the teacher was done, him and I walked back home. In the way back he mentioned neither the report nor what the teacher had said. In fact, he never revisited the subject.

One of my favorite pastimes as a child was to fly kites. When I was about five, one early spring afternoon, my mother dressed me up in a new light-grey gabardine tailor-made suit recently purchased for us by our maternal grandfather, Francisco. New, specially made boots, were also included. The occasion was an invitation for afternoon tea ("onces" in Chilean Spanish) at some of my mother's socialite acquaintances. As my mother began to ready my brother's clothes I went out to see if there was any wind to fly one of my kites. There was only a slight breeze so I needed the help of one of my friends, nicknamed Perejil (Parsley), to get the kite airborne. As my friend held the kite some 20 yards away I began to run backwards as he released the kite. This I had done countless times before but in this occasion I got too close to an irrigation channel and fell in... suit, boots, and all. When I came out I was soaking wet and covered in feathers. Undaunted, I asked my friend to help me get rid of the feathers while hoping for the September sun to dry me up. All was going well when my mother and brother came out all dressed up and ready to leave. My poor mother stared in disbelief as she observed the wet suit still with some feathers on it. I was immediately disciplined, the old fashion way, and given a change of clothes. Anyhow, we arrived for afternoon tea only slightly late.

During my early childhood my father worked in the aviation industry. He flew general aviation planes, was the navigator for a Liberator used mainly as a cargo plane, and ended up in charge of maintenance at the small airport southeast of Santiago known as Tobalaba. A self taught engineer who began to build radio transmitters at the age of eighteen he was highly independent. One day, in the early 1960s, he decided to go into the trucking business moving sand for the booming construction activity of the time. To this effect, he bought a Studebaker truck. It was hard work and he would get up at four in the morning. Mechanical parts, and spare tires, were kept in the patio of the house. My father's assistant was a Native American Chilean (of the Mapuche people) named Lisandro. A very strong man, his job was to unload the sand, side by side with my father, using large shovels. One afternoon they were back earlier than usual: Lisandro explained that my father had a "trucking duel" that ended up with the rival Chevrolet truck at the bottom of a river.

One day my brother and I discovered that getting into the empty tires, in a curl position, while other kids rolled the tire down the street was an exciting game. This game ended one morning when the tire with me as pilot, smashed into a group of shopping ladies and proceeded to cross a busy main street. That afternoon a group of people waited to complain to my mother and thus our wheel experiment came to an end.

Growing up in the western side of Santiago, Quinta Normal, in the 1960s was not without risks. From an early age we were exposed to street fights. The motive for a fight could be anything from possible cheating in a game of marbles to an insult. When I was about 8, I was confronted by a 12 year old kid, "a veteran," with a tough reputation. As the confrontation progressed I ended down on the floor and the kid kicked my face thus opening a cut on one of my eyebrows. I got up and the fight continued until some grownups intervened. As I was leaving, a man slowly applauded and said: "You didn't give up."

My maternal grandfather was a practical man from an aristocratic background; he was always well dressed, wore a necktie, and walked very upright. He hated vulgar language. In the 1920s he was among the first professional mechanics in the Chilean Army. As a civilian he kept a very nice mechanical workshop where he experimented on the conversion of mechanical energy into electricity. I had absolute freedom to play in the workshop when he was not there. All of the above introduced me to a set of circumstances dominated by the concepts of exploration and hard work.

Around 1964 we moved to the northern port city of Iquique. This was a result of my father's prosperous business selling Raytheon radar equipment, and giving maintenance to this equipment, that was installed in most of the fishing fleet. We stayed there for about two years. We lived on the peninsula of Cavancha that back then was a small fishing village. Initially, my brother and I had a hard time in school since we were considered as "southerners" and often we had to impose ourselves using boxing skills learned after football (soccer) games back in Santiago. Eventually, however, we were accepted. The beaches were pristine, the water clear, and we often went harpooning fish. These were happy times. On weekends my parents received guests and it was not rare to see several large lobsters being served for lunch. These lobsters were brought to my father as presents by grateful customers in the fishing fleet.

In Iquique, my mother was the public servant in charge of resolving labor disputes, an arbitrator, which took place mainly in the mining industry. Already the political climate was becoming charged and dominated by intrigue. She was ordered to move, by a new government, to a little agricultural town in central Chile called Peumo. Coincidentally, by this time, the fortunes of the fishing fleet, due to the temporal disappearance of anchovies, were dimming so the whole family moved back south.

Peumo was nothing short of paradise. It was a valley that produced fruits and wine. My brother and I became expert horsemen and in general we met great and lovely people. Here, our house was at the western end of the town next to one of Concha y Toro's vineyards. In fact our dog, a German Shepard called Niño, would get into Concha y Toro's vineyards and eat some of the small, and exceedingly sweet wine grapes, without apparent side effects. At school the girls tried to teach me "Cueca" (the Andalusian style national dance) so that I would be in the school's dance group. Albeit I did enjoy the company of the girls, I thought my legs were too stiff for dancing, and I rather preferred spending my free time playing soccer. One afternoon that I was due to play, I realized that I had left my soccer boots at home so I decided to go running to make it back in time. The distance I had to cover was about two miles each way, in a curved path, so I decided to take the straight alternative which was the rail line. This would mean entering the house from the back which was a field that we rented to an old farmer. Thus I began the run over the railroad jumping from sleeper (tie) to sleeper. There was a nice breeze blowing against my face from the west. As I was approaching the end of the run I saw in the distance a horseman who was signaling something. I thought he was trying to say hello waving one of his arms. As I got a bit closer I recognized the horseman as the old farmer we rented the field to. By now he had reached for his rope and began waving it side to side. All of the sudden, I felt a trembling on the ground and, without hesitation, proceeded to jump immediately off the railroad thus falling over wild blackberries that lined up the sides, never mind the thorns. As the Diesel locomotive went by I could see the engine driver reading a newspaper. The old man had saved my life but I had very little time to think about it. I had to get out of the blackberries, get the boots, and get back in time. I never told my parents, or anyone else, about this brush with death. Life in Peumo continued to be milk and honey.

It was now 1967 and we had to think about our education so we returned to Santiago where I applied to, and was accepted to, the *Instituto Nacional*. This all-boys school was the most prestigious high-school in the country and was founded by Chile's first president, Jose Miguel Carrera, back in 1813. Carrera was a warrior from the Napoleonic Wars. He participated in some twenty European battles and skirmishes against Napoleon's forces including the famous *Battle of Talavera* where he was wounded.

The school uniform consisted of a navy blue jacket, grey trousers, white shirt, blue tie, black belt, black shoes, and grey socks. Discipline was expected and enforced. Teachers were addressed as "Sir" and when a teacher entered the classroom everyone stood up. The first day at "The Instituto" I arrived early to the designated classroom, which was empty at the time. Thus, I proceeded to select a desk and chair at the left rear corner. Since this was the 8th grade I was transplanted into a cohesive class that proceeded from lower grades. I was one of a few new outsiders. As the students began to arrive they would tell me "Prat's desk" without further explanation. About ten minutes before the class was due to begin, Prat and a group of fellow students entered and, in a commanding tone, he proceeded to indicate that I had to vacate "his desk." Since I was not used to be spoken to in that manner soon a rapid fist fight broke out. The confrontation ended due to the intervention of several other students but already my beautiful white shirt and part of my jacket were profusely bloodied. It was Prat's blood. He was the great grandson of Arturo Prat, Chile's greatest naval hero, who died at the bloody Battle of Iquique in 1879, displaying, according to British historians, "outrageous courage." The contemporaneous Prat was also named Arturo. Perhaps in a genetic display of early manhood, and honor, he proceeded to take the desk next to me and never said a word to the head teacher.

In the late 1960s political trouble became apparent and life became politicized. This was particularly highlighted for us since my mother had become a high ranking career public servant, at the Ministry of Labor, and the government of the time (that called themselves Christian Democrats) gave her a hard time since she was not a member of the party. By then, my parents were spending a lot of time and money in lawyers fighting unfounded charges and my mother would refer to the situation as "political persecution." In 1970, further political upheaval came to Chile with the election of Salvador Allende who was a Marxist. Although he got about one third of the vote only, he was elected president with the parliamentary support from the Christian Democrats.

By now my father had his own business and among his customers was the US Embassy in Santiago, where he gave maintenance to their telecommunications equipment. My father, with his fierce independence, and a history of active anti totalitarianism during the 1940s, was reluctant to accept the idea of living under a socialist regime even though my mother was cousin of one of Allende's ministers. The situation was compounded by the following episode: one day, the contact my father had at the embassy, whose name was David Patterson, indicated that my father's name was on a "hit list" they had discovered. The list was said to be genuine and Patterson proceeded to recommended that we should leave the country. Albeit my father initially dismissed the idea (he was approaching 50), the prospects of us being left fatherless in a hostile political climate, eventually convinced him, and mother, that Patterson's suggestion should be taken seriously.

Leave, leave where? The US was the obvious destination. However, Patterson recommended that we should consider Australia as an alternative and provided some contacts. My father applied, and relatively quickly, after interviews and rigorous medical examinations, the family was accepted to leave for Australia. They even subsidized our air fares. In March 1971 we were on board a Pam-Am Boeing 747 from Los Angeles bound for Sydney. Thus we avoided the bloody events that unfolded in Chile during its 1972-1973 history.

Chapter 2

Sydney

2.1 Arrival to Sydney

In the way to Los Angeles, the Pam-Am Boeing 707 we boarded in Santiago, did stop in Panama, where the plane was given maintenance. When the plane took off, my father found in the toilet a wallet, with about one thousand dollars in it. He returned it to one of the Pam-Am stewardesses. As it turned out, the wallet did belong to the maintenance engineer stationed in Panama. In response, the stewardesses, with their wide white smiles, made every effort to reward my father and us with the best possible attention for the remainder of the flight. My brother and I then discovered that the English we had been taught at school "was not bad." Encouraged by this pleasant experience and a couple of little books entitled something like *Learn to Speak English in Seven Days* we went out shopping in Los Angeles, again with positive reinforcement. In the long Pam-Am Boeing 747 flight to Sydney, which made a stop in tropical Fiji, we practiced some more with a couple of English ladies who spoke to us slowly and very clearly.

We arrived in Sydney on a Sunday morning. It had been raining, but the clouds were dissipating and the sun was coming out. At the airport, immigration officials were waiting for us. We were soon taken to a waiting bus where other families were already seated. As the bus rode towards the migrant hostel in a place called Villawood,† what struck me most were the shining red roofs, made of red clay tiles, of endless rows of brick houses. Few apartments and buildings, mainly houses. This was a unique signature of Sydney at the time.

The hostel was a place where the Australians housed the new arrivals. There, they could stay for up to a year. Some of the housing was in former military installations: half cylinders. There was a waiting list for apartments. Food was served in large cafeterias three times a day. The food was good. It included a lot of lamb, chicken, baked beans, and even kangaroo meat. Soon we made good friends with other youngsters from Scotland, Ireland, and Uruguay. But not all was good. Within a couple of weeks my brother and I broke into a most annoying and severe skin rush, in our arms and legs, which eventually got infected and required treatment with antibiotics. My father took a job at the engineering workshops of the New South Wales Railways. My mother began to study English and soon she was capable of translating from French to English for some French families. It turns out that she had learned French as a youngster in school. We were taken to school and given a test based on some pattern recognitions and numerical tasks. Both of us were placed in the 11th grade at the Chester Hill High School.

2.2 Chester Hill High School

The principal was a kind gentleman named Mr. MacKerras. The uniform was gray trousers, white shirt, crimson tie, and crimson jacket. However, due to the warm climate, the jacket was optional. The school was of mixed gender and the population of students was largely Australian. I still remember the curiosity we evoked in some of the girls that would constantly stare. I never asked why.

As soon as we arrived at school we were given a bunch of books according to our selection of subjects. English was mandatory. We were also immediately renamed with English names: Henry and Frank. At this stage, after confronting the distinct Australian English rapidly spoken by our new teachers, it became absolutely clear that the English we knew was utterly insufficient. I began to read as much as possible and every Sunday evening watched an excellent English lesson directly from the BBC… "The Benny Hill Show."

We also took science, mathematics, art, and ancient history. Then, I signed in for a subject with the deceiving title of "industrial arts." This was divided into materials science, technical drawing, and mechanics. That is, mechanics as in physics and included the study of static forces and kinematics. The teacher was a Mr. Johnson, a spit and polish disciplinarian who saw his function as preparing his students for engineering studies. This was probably the best class I had ever attended. At the beginning I did not understand much of the materials science section as I became familiar with the phase diagrams of different alloys. However, soon I was scoring high in technical drawing and in mechanics. I remember I did every problem in the textbook and towards the end of the 12th year I was the only one to solve the most difficult problems in kinematics.

In mathematics we had several teachers. All of them excellent. Again, I mastered this subject well even though at the beginning I had to take the book with me to the soccer games on Sundays, where my brother was playing in the top youth squad of APIA, one of Sydney's first division clubs. One Sunday afternoon I taught myself differential calculus and the next Sunday integral calculus, I was floating in air.

One day the science teacher gave us some projects. I did mine on radio waves. The catch was that one had to give a brief talk on what was learned. Although by now I could read and write fairly well the pronunciation of my English was, as the Australians would say, "bloody awful." Soon after I began my presentation, the male Aussies started to smile nervously and then began to laugh... and laugh. The teacher and the female students, clearly embarrassed, did not laugh but had a strange expression on their red faces. Undaunted and undeterred, I continued as I noticed that tears were rolling down the faces of some in the audience. I did not stop until I completed my presentation. The apparent reason for this bizarre display of laughter was the use of the word "fact" ... which I repeatedly mispronounced to mean something else.

In English, art, and history, I shared classes with my brother. The art teacher, Mrs. Butterfield, took a special interest in the "Spanish boys" and was always asking if we needed help. She was very friendly and she was the only one who tried, and tried hard, to pronounce my name "Franciscoou." Mr. Kerry, the history teacher, went as far as allowing for on site translations of his dissertations. Needless to say that most of these translations, performed by a bilingual friend, were wasted talking about trivialities. As the High School Examination got closer, by the end of the 12th year, Mr. Danny Pace, the English teacher, wrote summaries of the books we were assigned to read. He did this in long hand over the blackboard. Two of the books were *The Tempest*, by Shakespeare, and *Death of a Salesman*, by Arthur Miller. The third was from an Australian author based on the story of a boy which evolved in the coastline of Perth. I think it was called *Merry-go-Round in the Sea*. With his transparent summaries the lecture of the books was facilitated greatly. As it turned out, essays from only two books were necessary to deal with the exam satisfactorily, so that was an enormous help. Our chances to enter a university were also improved by the availability of Spanish as a subject. Most of this was done through the mail and classes were only occasionally required. For that we traveled to down-town Sydney where less than twenty students, from all over the city, participated. Besides the study of grammar, most of the activity revolved around the lecture of classics authors such as Miguel de Cervantes. The teacher, a French lady, towards the end of the course said to me that I ranked among her best students ever. That was a welcome remark prior to the coming examinations.

Towards the middle of the12th year we moved to an old Sydney suburb called Strathfield where we rented an old brick house in Leicester Avenue. We would take a train to a little railway station called Leightonfield North. From there, we would walk about a mile to the school. In the afternoon we would run, books and all, back to the station to catch an early train. Missing that train meant waiting while the passage of a few cargo trains would rock the little station. I would look closely at train engines and heavily loaded cargo carriages and wonder how much fun would be to design them. On the other hand, my success in technical drawing and studies in art created a keen interest in architecture. But, what I enjoyed the most was the study of mechanics in both its forms: static and kinematics.

Despite serious cultural differences and communicational challenges, in this Anglo enclave, we were treated with tremendous deference, implied warmth, and fondness. Certainly we were different, Henry became the star of the soccer team and the team, in 1972, went farther than ever before in the regional championship. Some teachers also mentioned our courtesy and that we were considered a "likeable pair." Nevertheless, I think that what we experienced then was a genuine expression of that intangible quality called *fairness*... and for that, I will always be grateful.

That summer we waited for the examination results. By then, my father became ill and had to go for surgery. We went looking for work. On a Friday we saw a newspaper advertisement offering "good pay" for construction work. Next morning at about 6 am we showed up at, what turned out to be, a recently demolished site. The foreman looked at the group of about twenty individuals and pronounced with a firm voice of accomplishment: "Men, you are all hired." It was my first real job ever. We were all given a set of tools, including picks and shovels and were ordered to dig trenches on the red clay. I still remember the pick bouncing... and bouncing, with each blow. I still remember the torrent of sweat and the pain in my hands as I dug "my trench" under the unrelenting Sydney sun. Before noon, we were told to stop for lunch. Only then I realized that of the twenty or so hopefuls, only a handful of "men" were still at the job. The others had deserted. That afternoon we went home exhausted. But the pain and the exhaustion were no match for the pride we experienced.

For a week we went back day after day. Even tough we wore hats we ended up with open sores, on our necks, due to severe sun burns. The blisters on our hands had become a bloody mess. One of the men that had stayed in the job was a university student, a New Zealander. During lunch we would talk about history and politics which blended with some civility the harsh reality at hand. Our first employment ended one afternoon when I almost came to blows with the foreman for what I considered "use of vulgar language." What I did not realize at the time was that construction men spoke like that all the time. Despite the confrontation, the foreman paid in full, and in cash, what was owed to us. Soon after that, I found a job at a Woolworths warehouse and Henry in a factory of metal furniture.

One day the results of the examinations arrived. I was offered entry at the University of New England, at Armidale, and at Macquarie University. Architecture was taught at The University of Sydney, but given my performance in the English part of the exam, which allowed for a bare minimum pass, I was not offered a place there. The decision was easy. Macquarie was a train and bus ride away from home and it offered studies in electronic engineering which were of significant interest. With my father ill, and my mother studying English, it was my brother, by now working at the New South Wales Railways, and playing soccer, who assumed the financial burden of the home. Somehow, the family was able to put together some savings to pay for my first semester at Macquarie.

† Things have changed a lot in Australia since the 1970s. In recent years, Villawood has been in the news due to riots by illegal immigrants.

Chapter 3

Macquarie

3.1 Arrival to Macquarie

My first visit to Macquarie, one sunny afternoon, began with a train ride from Strathfield to Epping. There, I took a green public bus to the university. I was impressed by the beautiful soft green hills surrounding the building complex. A kind gentleman, in a grey uniform, was greeting the new arrivals and directing them to the place of enrollment. At the central courtyard, after contemplating the different options, without any doubt or hesitation I registered myself in physics. My advisor was Dr. John Robertson (D. Phil. Oxford), who suggested that besides all the physics and math, I should also try some chemistry.

The first day of class, March 1973, I had Physics 1A in the morning plus calculus. The first class in the afternoon was chemistry. The tutor wrote on the board a typical chemistry reaction. I asked him to please explain how he got from the left hand side to the right hand side. He replied: "This is the way it is." I immediately got up and went to see Dr. Robertson and explained to him that I could not do chemistry. He smiled and proceeded to suggest a biology course instead.

What I remember most of the early days at Macquarie was the physics professor. His name was C. E. Curnow. Claude Curnow was a near-albino and his eyes were of different pigmentation. He was of medium height, in his 60s, and always wore a grey overall with his pockets full of chalk. He spoke very clearly and his high pitch voice resonated supreme in the large lecture theater. In the theater there were about 9 central boards three sets of three. In addition, on each side of the central board there was a vertically revolving board. Professor Curnow believed in teaching physics by solving problems. In each lecture he would introduce the theory and then would begin solving problems. By the end of the hour each and every board was neatly written with physics. He gave us his own type written books which had scores of problems, from the most simple, to the most complex, solved in detail. The books were most original and they were written in such a way that the problems could be read without seeing the solution until the turning to the next page. White pages for theory, yellow pages for problems. Static mechanics, kinematics, Newtonian gravitation, name it, it was there. He was a true and awesome professor and I still have his books.

Professor Curnow taught the basics: mechanics and electricity. Then, in second year, we jumped into the *Feynman Lectures on Physics*. By now the lecturers were different and the class sizes had been drastically reduces to about 20 per class. The lectures where now: E. Brown, K. S. Imrie, R. E. B. Makinson, J. A. Robertson, and J. Unsworth. Arthur Pryor came from Lucas Heights to introduce us to nuclear physics. All the courses we took had a solid experimental course next to them. The laboratory work did not count towards credit but it was compulsory to complete the course satisfactorily. In charge of the labs and tutorials were A. J. Farrow and I. L. Guy. In addition to the physics there was a whole series of mathematics courses we took. Initially they revolved around algebra and calculus. The school was the combined *School of Mathematics and Physics* and the professor in charge, and calculus lecturer, was Professor Frederick Chong. Professor Chong was born in Australia, spoke perfect English, and was a captivating lecturer.

Third year physics was a different ball game altogether. To the previous lecturers we added G. C. Fletcher and J. C. Ward. Besides a distinct jump in demand for book work and theory third year physics had a unique added component. His name was Elmer Laisk. Elmer was from the University of Tartu in Lithuania. He was a small man (apparently in his late 60s) with white hair, a crazy benevolent look, and of course spoke with a heavy accent. As John Ward would recall later in his memoirs: "He had an encyclopedic knowledge of classical physics" (Ward, 2004). In a whole second floor of one of the buildings Elmer had created an experimental physics paradise for third year physicists.

The third year physics lab is something unlike anything I've ever seen. It was immaculately clean, it had beautiful blue benches, and the benches... were empty. Each bench had a name for a particular experiment and with each experiment there was a neatly bounded set of notes explaining the physics of each experiment. That's all. In the wall, at the back of the lab, there was a set of tall cabinets with all sorts of physics gear. The first exercise was to figure out how to assemble an experiment from scratch. All experiments were open ended and there was no time allocated to each of them but all had to be done by the end of the two semester courses suitable named as *Experimental Physics*. Among the experiments, I worked on, was the Frank-Hertz experiment, the making and testing of a germanium transistor, the Stern-Gerlach experiment, LC transmission lines, NMR, and the Hall effect. Elmer also had a CO_2 laser made in house.
The lab was attended by a quick witted Australian called Les Young Lee. Les was a technical wizard always willing to help and invariably with a friendly greeting. I still remember him coming back after lunch, smiling broadly, almost always talking about sheilas and asking: "G'day mate, are you getting any?"

The backbone of the third year physics curriculum, as designed by John Ward, were the following courses: Electromagnetism (Feynman *et al.* (1965); Lorrain and Corson, 1970), Quantum Physics (Feynman *et al.*, 1965), Solid-State Physics (Kittel, 1971), Experimental Physics III. In addition to these required courses I took Electronics II, Circuit Theory, Differential Equations, and Fourier Theory, Materials Science, and Undergraduate Seminar.

Professor Chong organized, at his home, every year a dinner for students who had distinguished themselves in Fourier Theory. I had the honor to be invited. It was a delicious and elegant dinner with many courses and I still remember the duck and the lobster. That was the very first time I had ever tasted Chinese food.

In Electronics II my lab partner was a quintessential Australian with a witty sense of humor called Ian Hill. Once a week we had to sit down with Dr. Robertson to explain our weekly assignment work. Ian would refer to Robertson as "John Robo" and constantly joke about his "excessive height." Confronted by Ian's witty and audacious remarks, Robertson would often say: "Stop dragging the chain, Ian."

Notice that in my curriculum there were no humanities, no English, no history, no economics, nothing of the sort. How could I do so much of the "hard stuff"? Simply. The Macquarie system was designed to be ultra liberal. At the time you were only required to take a structure leading to a major and the rest was up to you. So, though many used the system to get an ultra liberal education I used it the other way around and never took any of what we physicists considered "soft options." More about this later.

One other peculiarity of the Macquarie physics world, at the time, was that although you were required to do all your assignments and all your lab work nothing of this counted toward the grade. Everything depended on a final three hour written examination. I still remember vividly the prelude to my first physics exam, consumed by nervousness, with frequent trips to the toilet to get rid of every possible drop of water. As time progressed, exams became a welcome part of the Macquarie ritual. In fact, by third year I began to enjoy the feeling of euphoria associated with the rendition of a three hour exam. By then, I was so into it, that I would solve entire problems in my head whilst taking the train ride home.

At Macquarie physics, it was not easy to get high marks. In the first year I began scoring Cs which then began to improve into Bs. By third year I was achieving my best scores. In the original class there were some very gifted students who initially scored all As which were maintained into second year. They did this with little or no work. One of them did not even take lecture notes. The third year was so different, and the change in intensity so much, that they could not keep up and had to take a brake and come back to complete their degree later. In one second year math course, recommended for physicists (vector calculus), the exam questions turned out to be very different to what we had seen in class, or the assignments. Such was the difference that several of the best math students stood up and left in disgust. The physicist stayed. The three hour exam had six problems. I decided to divide the time and spend 30 minutes on each question. In all cases I ended up writing several pages of integral equations and did not get to a final answer. That night I took the train home thinking that I had met my first failure. I explained the situation to my father whom offered some words of reassurance. Dinner with my parents was a nice distraction. When the results came in, I discovered that I had passed. Not only that, later on, I was told that I had one of the best marks in the class. I had learned an important lesson.

For the first three years at Macquarie I was totally and entirely consumed by physics although I did take some summer jobs including a job loading frozen boxes of export beef, by hand, into container trucks at Sydney's historical Homebush abattoirs.

3.2 Physics Honours: metal-halide lasers

My original plan was to study electronics at Macquarie. However, the system required that prior to entering the engineering program a physics curriculum had to be completed. In doing so I had fallen in love with physics. Following the successful completion of my physics degree I faced the dilemma of what to do next. To this effect, I spoke to several of my professor who invariably pointed the way towards electronics. However, at the time a recent arrival from the University of Oxford had brought the possibility of doing research on metal vapor lasers. He was a New Zealander originally from Otago. His name was James Austin Piper.

At the time, we had built a successful family business building high-fidelity audio systems using British turntables (BSR) and speakers. My father had designed and built the amplifiers that were perfectly matched to the speakers. The sound was extraordinary, and we sold several systems per week as fast as we were able to make them, all by word of mouth. Things were going so well that we even had plans to expand into a little factory. Thus the issue of abandoning electronics was not an easy one. However, once I sat down with my father and I explained the dilemma his answer was quick and forthcoming: "Son, you must do what you like." As it turns out we kept on making hi-fi systems, in *DEE Electronics,* for a few years but eventually we had to stop given the demise of our suppliers due the Japanese competition.

The Honours program was an intense academic year in which the student time was divided between a demanding course load of upper echelon courses and a research project. My course work consisted in additional third year mathematics and advanced electronics taught by Ron Aitchison himself. These courses had a laboratory and a theoretical component. The two things that I remember the most from Aitchison's teaching were the following: first, "heat means death to a semiconductor," and secondly we derived from first principles, using Fourier theory, the *h* parameters that characterize transistors. One other thing, Ron Aitchison's exams were "open book" since he knew quite well that having the books there did not help and it was more of a diversion. To pass and to score well you really had to know the electronics inside out and the underlying physics. In his course, there were less than a handful of students and I worked in tandem with Brian Bulliman. When we would get tire of electronics, Brian and I would go out for a good game of squash and came back refreshed for more electronics.

My thesis project consisted on investigating the excitation cycle in metal-halide vapor lasers. At Oxford, working with C. E. Webb, Jim Piper had developed some unique metal-halide vapor lasers which emitted both in transitions from the metal and the halide atoms. Thus the emission was a spectacular array wavelengths and sometimes enough transitions occurred at the same time that white laser light was observed (Piper, 1977). Another advantage of the metal-halide approach is that the discharge of the laser could be operated at lower temperatures than if it were a pure metal vapor laser. Jim suspected that once the metal-halide molecules were dissociated, by electron impact, they would recombine to participate in the excitation cycle again. If they did not recombine, the atoms would adhere to the cold walls of the laser and eventually the discharge would run out of metal. To measure the diffusion rate of pure metal atoms and the recombination rate of metal-halide molecules Jim devised a clever double-pulse technique. A fast raising excitation pulse, about 10 microseconds in duration, would be followed by a sharp narrow pulse of about one microsecond in duration. The first pulse would dissociate the metal halide molecule and excite the metal laser transition. The second pulse would monitor, in the time domain, the intensity of the laser emission that would vary as the metal diffuses to the walls and/or recombines with the halide.

That was the idea. The technical details were left to me. Using what I had learned in electromagnetism about transmission lines, I proceeded to design and build a pulse forming network to excite the laser. This was an LC network using tailor made coils wound with a little machine in the electronics lab. My electronics training allowed me to design a high-voltage switch including a Philips thyristor. Since the load was a laser discharge it was non-Ohmic and matching the impedances was beyond the realm of calculations. It had to be done by hand. I clearly remember spending entire afternoons carefully modifying the coils in the LC network At the end, this routine got to be so tedious that I opted not to bother about the high voltage, whilst doing adjustments, and my fingers lost most of their little hair in the process. However, at the end I got a nearly perfect match… the current pulses were beautiful.

Without going into further details, I ended up measuring metal atom diffusion rates in a copper hollow-cathode discharge and molecular recombination rates in hollow-cathode metal-halide vapor lasers. The laser systems that I investigated were He-$ZnBr_2$, He-$ZnCl_2$, and He-ZnI_2. Understanding the data was not simple since, at the time, there were no published papers to compare results. To this end, I had long discussions with a physical chemist, Mike Batley, and with Dick Makinson. In the process, I became very familiar with the theory of chemical reactions and it was Dick the first to suggest that I was perhaps the first to measure the recombination rate of these molecules in a discharge (~ 50 □s for $ZnBr_2$ (Duarte and Piper, 1985)). Dick also said something else: "When you have a beautiful experimental result don't spoil it with some half-cooked theory."

The thesis was carefully written and highlighted the main result that the decay of metal species in the discharge was most likely due to recombination processes and that the molecular recombination rates did correlate with laser performance. The faster the rate the better the performance. A thesis entitled *Excitation Processes in Continuous Wave Gas-Metal Halide Vapour Lasers* (Duarte, 1977) was submitted. A talk was arranged for one afternoon in July, 1977. The talk was rehearsed several times and every conceivable question anticipated. I left home, at 1194 Pacific Highway, Cowan (Northern Sydney), around noon after having to push my battered Ford Falcon down the hill. The talk, I was told, went superbly. I was able to answers all the questions even with an air of authority. According to Jim Piper my thesis was good enough to be a Master's thesis. Soon after the talk the honours committee went into session.

It was around five that afternoon that a smiling Jim Piper gave me the news… I had been awarded *First Class Honours in Physics*. This was no small achievement. I had often been told that, in the British system, it was harder to get *a first* than to get a Ph. D. In addition, this was the first time that a first class honours had been given in physics at Macquarie. In practical terms, this meant a straight passage into the Ph.D. program whilst receiving a handsome scholarship from the Australian Federal Government.

Later, I learned that things did not go completely smoothly at the committee meeting. Although every one was impressed with the thesis work, and the presentation, there was some dissent from one of the mathematicians. In the face of this, I was told, Ron Aitchison, a quick witted, fast talking, and forceful character, who was also the Head of the School at the time, said something along the lines... "Frank is the best student I've had in the last ten years." Immediately the vote was taken and the degree awarded.

I am not sure if honours years in the British system continue to be as intense and high-pressure as mine was. But from my perspective this was one of the best, and most exciting, years of my life. I also never forgot Ron Aitchison's gesture, devoid of duplicity, devoid of mercurialness, truly representative of old-fashion fairness.

3.3 The science battles

3.3.1 The initial motive

During my honours year I began to spend a lot of time at Macquarie and subsequently made a lot of friends. A favorite thing was to go to the coffee shop and meet with some of my girl friends. There was also a bistro, a large cafeteria, and a bar. My social life began to expand. Most of my female friends were in the arts while most of my male associates were in the sciences. I also got involved in the Physics Club which was another excuse to meet girls during orientation by showing some very interesting exhibits that Elmer

Laisk had put together. It was in this social setting that the issue of Macquarie and its unique degree structure surfaced.

Briefly, Macquarie University was created as an ultra-liberal university whose degree structure was modeled after the old Oxbridge system. That is, one sole degree for everyone regardless of specialization or content of the degree. All that was required was for the student to complete a certain core set of courses, the major, and a certain distribution of courses at the various available levels.

Paradoxically, this enabled the students to either complete an ultra-liberal education, an ultra-scientific education, or anything in between. The one anomaly in this utopia was a newly created School of Law with its own named law degree.

Talking to physics and other science students it soon became clear that this was a major source of discontent and even resentment in the fact that they had to work "bloody hard" to get a science degree that did not even say that it was a science degree. On the other hand, our arts counterparts were having fun most of the time and ended up with the same named degree. Whether this could be substantiated quantitatively was another matter but this was the perception and perceptions can be powerful. In essence, the initial motivator was an issue of differentiation. However, we soon found out that the issue was far more complex than anticipated. Now, the issue of a named degree for the sciences, a B.Sc., was nothing new at Macquarie since Professor Chong had tried to introduce such a degree back in 1972. This idea was quickly rejected by the Academic Senate of the time and went dormant for the next few years.

3.3.2 The extended vision

Although the first B.Sc. proposal was easily defeated the players and the condition, by mid 1977, were vastly different. First, an unusual alliance began to develop between the science students and key members of the science professorial staff, notably between myself and John Ward. As a result of numerous informal conversations a vision beyond the purely differential began to emerge. This extended vision was key to sustain what eventually became a rebellion in the sciences and to gain the necessary support from external political figures. The elements of this vision were the following:

1. Qualified science students should have the freedom to choose between a BA and a B.Sc. This quest was, above all, about freedom of choice.

2. A named science degree was essential for Macquarie to compete, at an equal level, with other Australian universities in attracting science students and funding.

3. A named science degree would liberate Macquarie's full potential as a science power house.

This vision also implied that a named science degree would derail the unrealistic socialistic-egalitarian utopia installed by the dominant players of the education establishment. Most of the writings focused on the freedom of choice, on the disadvantages of the established system, and on the value that the change would bring to the university. We knew perfectly well the implications of our revolutionary vision and the opposition understood it quite well also. It was a classical confrontation of ideologies. John Ward, with a long history of confrontation with the educational establishment (Ward, 2004), would also see it as a battle between good and evil.

3.3.3 The conditions

Several conditions, political and otherwise, aligned in the 1977-1979 period which made the confrontation possible. The loss or absence of any of these components could have resulted in failure:

1. The presence and availability of science students willing to face the challenge. For instance, I had completed my Honours year and I had time on my hands.

2. The willingness of principal figures in the science faculty to support the students and their quest.

3. The bridges built between the science student body and the student population at large. Notable here, were the connections that I established with local center and mainstream student politicians in need of the science vote in local elections.

4. A long domination of student politics by theoretical far left-wing elements that alienated most of the student population. Although many of these radicals were nice people, their views were divergent with the practical interests of the average student.

5. The presence of a Liberal Party government in Canberra (Note: in Australia the Liberal Party is the "conservative party" and the Labor Party is to the left of the Liberal Party) and the bridges that I was able to establish with elements of that government.

6. An overly confident establishment.

7. The adoption of guerrilla tactics that allowed us always to be a step ahead of the status quo forces.

3.3.4 The opposing alliances

On one hand we had the science students organized into *Students for a Science Degree* (SSD). I was the president and other leaders included Milan Brandt, Brian Bulliman, and Michael Golding. These were the visible players. These were all math and physics but we also organized a network into chemistry, biology, and earth sciences. Our job was to organize the science student population into a disciplined voting block.

About the SSD: this was a very well thought out organization which beyond its internal representation had links to relevant professional institutes including The Australian Institute of Physics, The Royal Australian Institute of Chemistry, and The Australian Institute of Exploration and Geophysics. In all, SSD had more than twenty positions in its hierarchy. However, key to our successful tactics was that decisions were quickly taken at the top and acted on immediately or as soon as deemed necessary. Consultations with relevant members took place on an informal basis and meetings were rarely held.

Most science academics did support us either explicitly or implicitly. That support manifested itself, for instance, by allowing me to speak to the students for a few minutes prior to their lectures. This support was reinforced by the fact that visible professors throughout the sciences did openly support the movement. These included Ron Aitchison, Fred Chong, Brian Gray, John Hawke, Elmer Laisk, Dick Makinson, R. H. Vernon, and of course John Ward. Our support was strongest in math and physics and weakest in biology. Chemistry, with Professor Brian Grey, was our strongest allied.

At first, we sought support from the dominating left wing student politicians. After an initial friendly response the status quo leaders decided not to support us. Given the political domination of the far left, moderate student politicians belonging to the main stream political parties, the Labor Party and the Liberal Party, were in need of votes. Initially the Liberal Party was too weak and we formed an alliance with the mainstream-left labor politicians: the science vote for their support of the B.Sc.

Eventually, as the political map on campus changed, we shifted our support to the Liberal Party and a center-right Christian movement.

The *Moderate Student Alliance* (MSA) was born with our support. Allied student leaders in the MSA included Jeffrey Davidson, Geoffrey Provest, and Joe Moore. In addition to this inner campus political network I was able to link with activist of the New South Wales Liberal Party who provided valuable access to the Federal Liberal Government, led by Prime Minister Malcolm Fraser. Among my associates there was an interesting character who was an excellent orator. Very "conservative" he did not drink or smoke. However, he did have a weakness for ladies. One lunch time we were seated at the cafeteria and he arrived in the company of a nice looking lady who was a part-time student. By now, Macquarie was well-known to accommodate part-time students whom, in a significant proportion, were housewives of wealthy professionals coming back to complete their education. Right there, in public, he made a pass to this lady whom looked at him with some kind of motherly love and replied: "You wouldn't last me a minute."

It was never clear who the opposition was. They had no clear leader. They did however have sporadic spokesmen personified in senior professors from the humanities, notably education, and an Earth Sciences professor, specialized in geography, called A. J. Rose. I mention him because he was one of the few high-ranking status quo academics, opposed to our endeavor, that I actually met one-on-one to discuss our differences. The meeting was amicable, and civilized, but fruitless. There was also passive opposition. Remarkably, a member of this opposition was Peter Mason, Professor of Physics. I knew Peter well as he taught me sections of biophysics and materials science. Although officially his position was one of neutrality we considered him as part of the opposition since he was supposed to be one of us. Peter and I got along very well and we had an excellent rapport in regard to general matters but we rarely discussed the B.Sc. issue. Indeed he was a generalist interested in philosophy, interdisciplinary education, and he was very good at it. Albeit his contribution to the university was admirable, his effect on physics at Macquarie was somewhat subdued. Perhaps this was due to the enormous influence of John Ward. This opinion was openly expressed following an article written by Arthur W. Pryor in *The Australian Physicist* (Duarte, 1988).

As amorphous as the status quo opposition was, we knew they had the votes and the power, where it counted, in the Academic Senate. In retrospect, their biggest mistake was to procrastinate and not to take a vote early. This allowed us time to grow stronger and build up the pressure on the administration. In all this, the talented politician was the Vice-Chancellor, Emeritus Professor Edwin C. Webb, he played the role of the ultimate fence sitter. We surmised that he knew very well the strength of the opposition and did not dare to make a move for as long as there was uncertainty.

In addition to the intellectual opposition from some of the humanities professors, who saw their egalitarian utopia threatened, all the far left wing student activists assumed a similar position. We knew that if they held the student government we would surely be defeated.

3.3.5 The battles

The battle itself began modestly with passing skirmishes in Macquarie's *University News*. The tactic was simple: create and maintain pressure on the university establishment via a series of letters. Duarte, Golding, and Bulliman (1977) were the first authors in a series that included the occasional participation from members from the physics staff such as Trevor Tansley. This activity began in June 1977. At the same time we began a campaign to collect signatures from the students in a petition asking the Vice Chancellor to introduce a B. Sc. At the time there were some 1800 science students at Macquarie. Our goal was to collect some 2000 signatures. To this effect, we would talk to science students prior to their lectures and continuously walk around with a petition in places like the library, the science building courtyard (E7B), and the cafeteria. It was not automatic to obtain a signature and it often took a small conversation, and lots of smiles, especially with the girls.

The petition simply read: "Any candidate registered in one of the science schools who has satisfied the requirements for the award of a BA degree… and has completed the program of studies… with at least an aggregate of 42 credit points … in the above named schools shall be awarded a B.Sc. in lieu of a BA"

A clear escalation came in November of that year when the signatures were collected and delivered to Vice-Chancellor Webb. I contacted the *Sydney Morning Herald,* Sydney's major newspaper, and on the 9th of November, 1977, they published an article entitled "2000 seek introduction of science degree at Macquarie" (Parfitt, 1977). The article quoted both Professor Webb and I. Simultaneously the local newspaper, *The Northern District Times* (1977), run the front page headline "UNI REVOLT SIMMERS: Students campaign for a science degree." The article included my smiling face and some quotes including: "We are confident of victory... not for us, but for the sciences and Macquarie."

The petition plus the publicity set in motion the bureaucratic wheels. Internally, at the School of Mathematics and Physics, a working party on the B.Sc. was established. This was integrated by Professor Chong, Professor Ward, Associate Professor Makinson, and Frank Duarte. This working party took the unusual position in one of its recommendations to call on Macquarie schools which were not bothered by the lack of a B.Sc. "… not to stand in the way of science-oriented schools..." Externally the Academic Senate of the university established a *Science Committee* with a litany of tasks to look into. One of these tasks (number 6) read "That the Science Committee… be asked to report on the possible introduction of a B.Sc. degree." This committee had more than twenty members. The School of Mathematics and Physics selected Professor Aitchison and I, as representatives.

At the same time SSD continued to extend its work with external organizations and data gathering in support of its goal. This included a survey of the preferences of local high-school students and securing public resolutions of support from The Australian Institute of Physics, The Royal Australian Institute of Chemistry, and The Australian Institute of Exploration and Geophysics.

Alongside the bureaucratic struggle an open battle began for the hearts and minds of the student body. As mentioned previously, our first alliance was with the labor politicians. The verbal agreement we had was that we would deliver the science vote in exchange for their open and unreserved support for the B.Sc. In addition, we began to author a relentless series of articles in the student newspaper *Arena*. These rebellious articles were designed to be visually attractive, provocative, bordering on the insolent. The first such article was entitled *A what from where?* and made repeated sharp references to a previous deputy Vice-Chancellor responsible for shelving the first B.Sc. proposal in 1972. The article concluded with the question "Is Macquarie going to adjust itself to the present times or continue to pose as the New Oxford University?" (Bulliman and Duarte, 1977). Additional articles in the series included (Duarte, 1978a; Duarte, 1978b; Duarte and Brant, 1978). Also, a famous little poster reading, in red, *Stop the racket… a B.Sc. for Macquarie!* was seen all over the Macquarie campus.

The first student election, in September 1978, was a long and hard confrontation, but with the help of a disciplined science vote the radical left was swept away from office. That sent a loud and clear message to the administration and the rest of the university. The student body was with us. We also run successful campaigns to elect student representatives to the University Council that added to the pressure. The labor politicians did deliver on their promise but reluctances began to emerge and we allied with an incipient center right coalition organized around what became to be known as the *Moderate Student Alliance* (MSA). By now in an unprecedented political scenario the far left was almost completely vanquish from the Macquarie political scene that became dominated by the center left and the right leaning alliance.

By mid 1979 it was clear that the Science Committee would recommend the adoption of a B.Sc. degree, we had total control of the student body and the campus press, the professional institutes had expressed written support to our cause, I had personally brought the issue to the attention of the minister of education, Senator John L. Carrick, and John Ward had given press interviews acidly critical of the status quo. Obviously, by now the incentives for the Vice-Chancellor to come off his precious fence were rather obvious.

3.3.6 The external influence

My mother, Ruth Virginia, was the first South American Social Worker to have her professional title (from the University of Chile) validated in Australia, around 1972. As such she began to work with charities and Catholic Immigration in Sydney. As is often the case, in Western societies, this type of activity involved the wealthy and the socially connected. In Sydney, the wealthy meant people connected to the Australian Liberal Party. As such, my parents were often invited to social events and sometimes my brother and I would go along. It was in this type of setting that in one occasion I met the former Australian Prime Minister Sir William McMahon, and his wife Sonia, in Strathfield where we lived prior to buying a property in Cowan (north of Sydney near the scenic Hawkesbury River).

As time evolved, my mother developed contacts with the legal profession in Sydney which included two young lawyers one of whom was Geoffrey B. Ferrow. Geoffrey was a QC and a brilliant barrister who was also involved in the organizational structure of the New South Wales Liberal Party. In his early thirties he was very wealthy and an "archconservative." Of Scottish descent, his family had a property in the Blue Mountains and he lived in a house on the north shore of Sydney's harbour. His house was almost directly opposite to the Opera House. At the entrance hall of his residence there was a very large portrait of Her Majesty Queen Elizabeth II. A superb orator, Geoffrey was both admired and feared by the political establishment given his conservative outlook on life. It was through Geoffrey and his comrades that I was connected to members of parliament and eventually to key players of the Fraser government including Senator John L. Carrick, minister of education, who subsequently became Sir John L. Carrick, minister for national development and energy. Also, as consequence of my involvement with the Liberal Party I also opened contacts and corresponded with a local member of parliament, named John Howard, who would eventually become one of Australia's longest serving prime ministers.

In his memoirs, John Ward makes reference to my political contacts (Ward, 2004). I never explained to him, or any one else, how these contacts came about. It is fair to say that at any one time during this period I was the only one who knew all the protagonists involved and how all the connections fit together.

Often I had long conversations with Geoffrey and often these conversations would make reference to geopolitics, Marxism, and the observed decadence of Western society. One of those conversations took place one Saturday morning in the summer of 1980 at Manly Beach. We got there about 10 in the morning and began to walk up and down the beach. Just after noon we decided to go for lunch but prior we went for a swim. Sydney beaches were already topless territory and as we began leaving, a most impressive young member of the female species was lying over the golden sand, her beautiful face and perfect front anatomy uncovered, facing the glorious sun. Geoffrey, the archconservative, immediately reacted forcefully and said in a loud and grave voice:

"In the name of God and country… let me cover her… with my own bare hands!"

Geoffrey's sentence I never forgot and I reproduced it here verbatim. Once I came to the US I corresponded with him until one day the correspondence ceased. I went back to Australia, on a physics assignment in 1994, and I met with Geoffrey's law colleagues. When I enquired about him, they told me that Geoffrey had passed away. Apparently he was allergic, something that none of us knew, and after having dinner in a restaurant he had chocked to death. A sad end for a brilliant mind. Being a natural born skeptic I still wonder if Geoffrey's open anti-Soviet activities had any thing to do with his untimely oblivion.

3.3.7 Victorious

By September 1979 my first physics paper, with Joe Unsworth, a physics Ph.D. from Manchester, was due to be published. In a course called Materials Science, Joe had set up an experiment on thermal diffusivity. The physics to explain the temperature, as a function of time, in a rubber sphere was linked to a classic book by Carslaw and Jaeger (1973). The mathematics was based on Fourier theory and soon I discovered that the equation being used, from the book, did not apply to the problem at hand. Thus, I sat down to solve the problem from first principles. Soon I wrote down the right solution using what is known as Duhamel's Theorem but in the paper we opted to write the solution using a traditional Fourier approach. The theory was found to fit the experimental temperature profile, as a function of time, rather beautifully (Unsworth and Duarte, 1979). This work was done prior to the start of my doctoral research. Now I was hard at work on optically pumped lasers and working on what eventually would be a series of new multiple-prism grating, high-resolution, tunable laser oscillators.

By now, we had control of the political landscape at Macquarie and the pressure to give speeches and engage in inflammatory rebellious rhetoric had subsided. Nevertheless, I was in close contact with the leadership of MSA and I was even running for elections as an

Australian Union of Students representative. The political machinery we had created was fully engaged and cruising along. This was the atmosphere late morning on Tuesday, the 11th of September, 1979, when Ron Aitchison waved to me from one side of the Aztec style administration building. That morning we all knew that the Academic Senate was due to meet and that the B. Sc. figured prominently in the agenda. Ron had just left the meeting. We began walking towards each other sorting the gum trees in the main court yard of Macquarie. Ron was smiling broadly and as we approached he extended his right hand to perform that unique male ritual of a firm and sincere hand shake. "The B.Sc. was approved" is all he said. I did not say much either.

At lunch, an impromptu gathering of physics professors and students joined at a large table in the cafeteria. Most of the main players were there. We were smiling and we were content but there was no boasting and no gloating. After more than two years, that long promised victory had come and we were piercing into a whole new future for Macquarie.

A couple of days latter the *Sydney Morning Herald* run an article (Parfitt, 1979) with an opening paragraph that read "After more than two years of intense debate... Macquarie University has approved the introduction of a ...B.Sc. degree." The same piece goes on to mention that the final vote was 16-1 in favor if the introduction of the new degree and quoted me highlighting the new added freedom for Macquarie's science students. An article on the following issue of *The Bulletin* (September 25, 1979) mentioned the topic describing it as following some "particular fierce internal politicking" and added that some of the opponents "claimed they had been virtually blackmailed into accepting the proposal." An article by Sheridan (1980) in the same magazine refers to the B.Sc. campaign as a "nasty, bitter, bureaucratic struggle" and goes to highlight John Ward's harsh criticisms towards the education establishment.

The effect felt in sectors of the humanities was long lasting. Even in 1992 an official, or establishment leaning, version of Macquarie's history referred to Greg Sheridan as a "mouthpiece" and lamented the criticisms of Brian Gray and John Ward making particular reference to Ward's use of "scandalous and disgraceful" (Mansfield and Hutchinson, 1992). Their description of events continues: "The venom of the attack... formed a part of the new wave of activism... of science students and staff alike... Never one with patience either for due process or the social sciences, Ward was vocal in his denunciation... Suitably, then, it was a close student associate of Ward's... Frank Duarte, who began to mobilize student opinion in favor of a change" (Mansfield and Hutchinson, 1992). The same authors go on to accept that the "shockwaves" of the B.Sc. "ran far into the fabric of the university."

In his memoirs, John Ward does refer to this particular incident and writes: "I explained that I had not said that the university was disgraceful and scandalous. I had said that the actions of the education establishment, in building huge education and psychology departments by using their influence, were indeed so, which is an entirely different matter" (Ward, 2004). Further, for the record, as previously stated, my contacts with the opposition were extremely limited and always amicable. As far as I know, there was no attempt whatsoever to intimidate. If they felt intimidated, by the plethora of measures that we applied, on the establishment and institutions of the status quo, that is a different matter.

What our opponents failed to understand is that we acted in what we thought were the best interests of the university and the students. John Ward wrote about his early vision of Macquarie: *When I arrived at Macquarie it was possible to imagine the future institution as a possible equal to all but the great universities of the world* (Ward, 2004). John did not articulate this vision to us at the time but thoughts along these lines were already implicit in our ethos (see 3.3.2). It was that vision coupled with the vision of freedom of choice for the science student population that motivated the science's revolt. Our revolution had an irreversible effect on the sciences, and Macquarie as a whole, *that was our victory.*

3.3.8 The AUS incident

By the end of 1979 we had a full political locomotive which was unstoppable. Planning for a longer and harsher confrontation I became part of the MSA ticket for the Australian Union of Students. This was the national body that integrated all the student university governments in Australia. Its importance was that it controlled a sizable budget which was spent according to its yearly political agenda.

Following our victory at Macquarie the real motive for this extra step evaporated but I decided to go on anyhow. Our ticket included Joe Moore, Jeff Davidson, Mary de Andrade, Geoff Provest, and myself. The elections came and I got elected and a group from Macquarie headed for Monash University. From the labor ticket Steve Magee, an old allied, was also elected. To understand what I am going to write next the reader should understand that AUS was so far left that the normal left, that is, the Labor Party, was considered to be right wing. Those to the center and right of labor were considered to be extreme right. The Macquarie revolution was unique in Australia, at the time, and AUS continued to be in the firm grip of the radical left. Some dominant characters in AUS appeared to be Marxists-Leninists of the pro-Soviet line.

Almost immediately at Monash I realized that the whole thing was going to be a waste of time, so I decided to limit myself just to have fun. A good thing about the "right-wing" delegation was a healthy quota of good looking girls which made the whole thing rather palatable. The exercise consisted in the far left pushing a whole set of radical motions and the opposition trying to put the brakes. The bad thing was that many of these motions had a purse attached to them.

On Saturday night a large group of us decided to go downtown Melbourne for dinner. After a few hours in the town we came back about 2 in the morning to witness the diehard leftists pushing their agenda. I went to bed and got up about 8 am and went for breakfast.

In my way there, I noticed that the same session was still going on. After a nice breakfast I went back to the session and saw Steve Magee still stoically defending his views before the overwhelming left-wing establishment machine. I should mention here that the language used by many of the delegates was beyond colorful and often obscene. By now, the session was nearly full of delegates. I stood up and asked the president to allow me to speak. I was standing at the right end of a big U table arrangement. Then, I began to castigate the audience using perfectly clean language. Perhaps too clean, because the horrified president ordered me to cease my speech at once. When I kept on going my microphone was turned off. At that time, using a loud voice I began an interchange with the president and I was threatened with expulsion. Undaunted, I jumped over the U table and began to run towards the president. I still remember running over the tables like if in slow motion... and as I got near the president I remember a face of astonishment of one of the delegates who was about to rest her cup of coffee over the table. At that moment I kicked the cup of coffee and the liquid caffeine bathed the comrade president. Fortunately, the cup missed him.

Immediately a vote was taken and I was summarily expelled from AUS. The first, and perhaps only individual to experience such "punishment." I flew back to Sydney with Steven Pringle. Right away, I went to see my barrister friend Geoffrey Ferrow, at his elegant office in Martin Place, who quickly determined that I, as an elected representative, could not be expelled. Thus, I returned to

Melbourne with a letter. AUS immediately recognized their mistake and proceeded to reintegrate me and to pay all my travel expenses. I was given a heroes welcome by my "rightists" friends and I earned even smiles from the leftists who would say "welcome back" and others "are you going to behave now?" Later I read a written account of the incident in a student's newspaper that described a "mad Chilean rightist goose-stepping over the tables."

The "goose-stepping" exaggeration reminds me of John Cleese and another physics friend called John Liggins. John was an extremely tall, bright, and eccentric character. Somehow, at about this time, he and a group of friends ended up with some kind of grant to film a movie. As the movie was being filmed John asked me to appear in it. My appearance would exploit the large size of my moustache, he said. So I did. Not even at the screening of the movie I knew what the movie was about and he never bothered to explain it to me. All I remember is my appearance… as a Mexican outlaw.

In 1980 I was also elected member of Macquarie University Students' Council, a position which I took somewhat lightly as I focused more and more in my doctoral research. By then, I had already authored my first paper on high-resolution tunable lasers (Duarte and Piper, 1980) and I was working on what would become a new class of tunable laser oscillators (Duarte and Piper, 1981). Nevertheless, I somehow always found time to continue the political discourse.

As a footnote it should be said that my set of beliefs did not prevent me from developing friendships with political adversaries. During my political deals I became involved with a young lady well-known in left-wing political circles at Macquarie. Thus, it was a little bit of a surprise, among friends and foes, when following a trip to South America I showed up with a young Chilean wife on campus. Lee Kirman, who was a very bright math and physics postgrad, summed up this unexpected development with the adjective "stealth."

References

Bulliman, B., and Duarte, F. J. (1977). A what from where? *Arena* **10** (7), 10 .

Carslaw, H. S. and Jaeger, J. C. (1973). *Conduction of Heat in Solids.* Oxford, London.

Duarte, F. J. (1977). *Excitation Processes in Continuous Wave Gas-Metal Halide Vapour Lasers.* Honours Thesis, Macquarie University, Sydney.

Duarte, F. J., Golding, M., Bulliman, B. (1977). Science degree. *University News* **1** (100), 16.

Duarte, F. J. (1978a). What's been happening with the B. Sc., *Arena* **11** (11), 9.

Duarte, F. J. (1978b). The Macquarie outlook, *Arena* **11** (13), 5.

Duarte, F. J., and Brandt, M. (1978). B.Sc. at Macquarie? *Arena* **11** (9), 3.

Duarte, F. J., and Piper, J. A. (1980). A double-prism beam expander for pulsed dye lasers, *Opt. Commun.* **35**, 100-104.

Duarte, F. J., and Piper, J. A. (1981). Prism preexpanded grazing-incidence grating cavity for pulsed dye lasers, *Appl. Opt.* **21**, 2113-2116.

Duarte, F. J., and Piper, J. A. (1985). Measurements of neutral metal decay rates in CW rare-gas metal-halide vapor lasers. In *Proceedings of the International Conference on Lasers' 84* STS Press, McLean, VA, pp. 261-267.

Duarte, F. J. (1988). Peter Mason and Physics at Macquarie. *Australian Physicist* **25**, 52.

Mansfield, B., and M. Hutchinson, M. (1992). *Liberality of Opportunity: A History of Macquarie University 1964-1989,* Hale and Iremonger, Sydney.

Parfitt, C. (1977). 2000 seek introduction of science degree at Macquarie. *The Sydney Morning Herald,* November 9.

Parfitt, C. (1979). Macquarie University approves B.Sc. degree. *The Sydney Morning Herald*, September 13.

Piper, J. A. (1976). Simultaneous CW laser oscillation on transitions of Cd^+ and I^+ in a hollow-cathode He-CdI_2 discharge. *Opt. Commun.* 19, 189-192.

Sheridan, G. (1980). Australian physicist wins Guthrie medal. *The Bulletin* **101** (5239), 49-50.

TNDT (1977) Uni revolt simmers: students campaign for a science degree. *The Northern District Times,* November 9.

Unsworth, J., and Duarte, F. J. (1979). Heat diffusion in a solid sphere and Fourier theory. *Am. J. Phys.* **47**, 981-983.

Ward, J. C. (2004). *Memoirs of a theoretical physicist.* Optics Journal, Rochester.

3.4 Appendix I: A brief history of the BSc at Macquarie University†

One day in 1977 a few students, of the *School of Mathematics and Physics*, gathered at the courtyard of building E7B to discuss the academic structure of Macquarie University and the implications of such structure for those who decided to follow a scientific career. A predominant original thought among those present was that although someone could complete an ultra rigorous scientific program, allowed by the Macquarie system, at graduation there would be no differentiation with students who elected a liberal path of education. This was a consequence of the Oxbridge tradition adopted by the founders of the university. A tradition that according to those gathered was not relevant to the Australian market place where science graduates were expected to have science degrees. Hence, the group decided to pursue the quest for a BSc degree.

The student movement was organized into STUDENTS FOR A SCIENCE DEGREE (SSD). The chairman of SSD was Frank Duarte and the leadership included Milan Brandt, Michael Golding, and Brian Bulliman.

Beyond the issue of the name of the degree it was understood early in the movement that such a change in the degree structure would have a significant and profound effect in the future direction of Macquarie University. Indeed, it was apparent to the strategists that Macquarie needed such change in order to enhance its science student population and to augment its scientific reputation in the community. Thus, this incipient student movement soon gained the strong support of leading academics within the sciences. Among the professors supporting the science movement were J. C. Ward, R. E.Aitchison, R. E. B. Makinson, F. Chong, E. Laisk, J. Hawke, B. Gray, and R. H. Vernon.

The BSc campaign lasted about two intense years and was described in an article in *The Bulletin* as a "nasty, bitter, bureaucratic struggle." It involved a direct, pervasive, and effective student campaign that included the participation in university wide elections, and the achievement of victory in these elections in association with the Moderate Student Alliance (MSA), the consultation with federal government ministers, the support of professional institutions (such as the Australian Institute of Physics), and a well coordinated publicity campaign.

On the 11th of September of 1979 the Academic Senate of Macquarie University approved the introduction of the BSc degree thus for ever changing the destiny of the university.

In retrospect, it is fair to say that, the confrontation of ideas that led to the introduction of the BSc yielded a more versatile and stronger Macquarie University. Further, it enhanced the career opportunities for a significant fraction of the student population. In this regard, this might be a singular example of a successful and lasting contribution exerted by student political action.

Much of the history of the BSc campaign was documented in *Arena* in the 1977-1980 period. In addition to the issue at hand, these articles provide a rare perspective into the realm of student and university politics of the times.

Selected Bibliography

Uni revolt simmers, *Northern District Times*, 9th of November, 1977.

2000 seek introduction of science degree at Macquarie, *The Sydney Morning Herald*, 9th of November, 1977.

F. Duarte and M. Brandt, BSc at Macquarie?, *Arena* **11**(9), 3 (1978).

F. Duarte, What's been happening with the BSc, *Arena* **11**(11), 9 (1978).

F. Duarte, The Macquarie outlook, *Arena* **11**(13), 5 (1978).

F. Duarte, AUS the redundant singularity, *Arena* **12**(1), 12 (1979).

C. Parfitt, Macquarie University approves BSc degree, *The Sydney Morning Herald*, 13th of September, 1979.

G. Sheridan, Australian physicist wins Guthrie Medal, *The Bulletin* **101**(5239), 49-50 (1980).

B. Mansfield and M. Hutchinson, *Liberality of Opportunity: A History of Macquarie University* (Macquarie University, Sydney, 1992).

†As published on http://www.opticsjournal.com/BSc.htm on 13th of April, 1998, and last modified on the 26th of March, 2005.

3.5 Appendix II: Macquarie Science Quotations†

"A far more serious problem is the one related to the future of Macquarie University as a science institution. In this respect one could quite legitimately ask: will science survive at this University if the present trend continues?" (F. J. Duarte, M. Golding, and B. Bulliman, *University News*, June 1977).

"Is Macquarie going to adjust itself to the present times or continue to pose as the New Oxford University?" (B. Bulliman and F. Duarte, *Arena*, July 1977).

"It is hoped that the action of our own students will contribute to the aim of letting science at Macquarie to be seen to be done! (T. L. Tansley, *University News*, August 1977).

"Uni revolt simmers... students campaign for science degree" (*Northern District Times*, November 9, 1977).

"We are confident of victory" (Frank Duarte, in *Northern District Times*, November 9, 1977).

"Indeed, it is quite obvious that the present situation at Macquarie is analogous to that of being in a tunel with only one clear end" (F. J. Duarte and M. Brandt, *Arena*, August 1978).

"Introducing a BSc would not only add prestige to the University but also longevity. It would not be a deceptive and fraudulent expense... It will assist in promoting the University as a balanced functional universal institution" (F. J. Duarte and M. Brandt, *Arena*, August 1978).

"The last edition of *Arena*, headlining the BSc issue, was banned from being distributed... other articles... written by 'progresive left'... were not called into question..." (*Arena*, September 1978).

"... for the first time in the history of Macquarie University an issue inherent to our very own academic environment became the core of the political process" (F. J. Duarte, *Arena*, September 1978).

"... the great majority calls for a change from the present structure, a change that will lead towards a better and stronger Macquarie" (F. J. Duarte, *Arena*, September 1978).

"After more than two years of intense debate, the academic senate of Macquarie University has approved the introduction of a... Bachelor of Science degree" (*The Sydney Morning Herald*, September 13, 1979).

"The decision represents the fruits of a two-year campaign... and follows some particularly fierce internal politicking at the university" (*The Bulletin*, 100 (9), 140 (1979)).

"...some of the oponents of the degree... complained of *naked intimidation* by the science people" (*The Bulletin* 100 (9), 140 (1979)).

"I was one of the noisy *BSc pushers*, perhaps, I was the noisiest of the *BSc pushers*..." (F. J. Duarte, *University News*, September 1980).

"If there exists any kind of division in the Macquarie academic community, it is not between left and right, nor is it between the Scienes and the Humanities; it is between those who believe in academic excellence and those who don't" (F. J. Duarte, *University News*, September 1980).

"One of *Ward's* few close friends at Macquarie is... Frank Duarte... the two make an odd couple - the restrained rather distant Englishman and the intense, earnest South American" (*The Bulletin* 101 (11), 49-50 (1980)).

†As published on http://www.opticsjournal.com/macquariephysicists.htm on 6th of March, 2005, and last modified on the 11th of March, 2005.

3.6 Appendix III: School of Mathematics and Physics

Incomplete listing of academic staff at the School of Mathematics and Physics in the mid to late 1970s:

R. E. Aitchison (Electronics)
R. J. Andrews (Mathematics)
R. D. Bell (Mathematics)
R. A. Brown (Physics)
P. G. Browne (Physics)
F. Chong (Mathematics)
B. M. N. Clarke (Mathematics)
C. D. H. Cooper (Mathematics)
J. V. Corbett (Mathematics)
C. E. Curnow (Physics)
J. L. Drummond (Mathematics)
M. J. Edgeloe (Mathematics)
A. J. Farrow (Physics)
G. C. Fletcher (Physics)
T. E. Freeman (Physics)
I. L. Guy (Physics)
K. S. Imrie (Electronics)
E. Laisk (Physics)
R. E. B. Makinson (Physics)
P. Mason (Physics)
J. E. Moyal (Mathematics)
A. G. R. McIntosh (Mathematics)
D. A. Payne (Mathematics)
B. G. Pillans (Mathematics)
J. A. Piper (Physics)
A. W. Pryor (Physics)
E. D. O'Keefe (Mathematics)
R. A. Robertson (Physics)

G. H. Rossmanith (Mathematics)
R. H. Street (Mathematics)
T. L. Tansley (Physics)
J. Unsworth (Physics)
A. J. van der Poorten (Mathematics).
A. E. Vaughan (Physics)
J. C. Ward (Physics)
T. Wright (Mathematics)

I took classes from those whose names appear in *italics*.

Chapter 4

John Clive Ward

4.1 Introduction

John Clive Ward was a modest individual with a shy temperament. In all the time that I interacted with him he rarely volunteered information about his physics accomplishments and only brought up aspects of these if relevant to the conversation at hand. Hence, a brief summary of his contributions is given here.

J. C. Ward is the creator of the celebrated *Ward Identiy,* also known as *Ward Identities,* which is a central piece of renormalization theory (Ward, 1950). An extension of this physics is known as the *Ward-Takahashi Identity.* This work was inspired by a brilliant conjecture of Dyson in a famous paper on the equivalence of the quantum electrodynamics theories of Feynman, Tomonaga, and Schwinger (Dyson, 1949). A recent book on the subject explains that the *Ward Identity* "ensures the universality of the electromagnetic interaction" (Greiner and Reinhardt, 2009). For this contribution Sakharov classified Ward as one of the *titans* of quantum electrodynamics alongside Dyson, Feynman, Schwinger, and Tomonaga (Sakharov, 1990).

Previously, using Dirac's notation, he was the first to work out the quantum polarization theory of two counter propagating photons (Pryce and Ward, 1947; Ward, 1949). He was also a main contributor to the *Standard Model* via papers he coauthored with A. Salam (Salam and Ward, 1961, 1964a, 1964b).

In addition, John collaborated with well-known mathematicians and theoreticians thus producing a series of brilliant papers on: the Ising Model (Kac and Ward, 1952; Montroll *et al.*, 1963), quantum solid-state physics (Ward and Wilks, 1952), quantum statistics (Montroll and Ward, 1958), and Fermion theory (Luttinger and Ward, 1960). His last paper was on the Dirac equation and higher symmetries (Ward, 1978).

In a piece written in an Oxford publication it was once stated that Ward "has contributed deeply to an astonishingly broad range of theoretical physics: statistical mechanics, plasma physics, quantum electrodynamics, and particle physics" and the writer continues "he has drawn attention to basic truths, and has laid down basic principles, which physicists have followed in subsequent decades, often without knowing it, and generally without quoting him." (Dunhill, 1995).

Dick Dalitz, another famed Australian physicist, wrote: "Ward was very capable at finding his way through complicated systems with many degrees of freedom, such as field theory and many-particle theory. His brilliance shone in research on such problems" (Dalitz and Duarte, 2000).

It should be noted that as of 2011 a Google search of "Ward Identity", and physics, yielded more than 77 thousand sites. Further accounts on the contributions of J. C. Ward can be found in Dalitz and Duarte (2000), Duarte and Piper (2000), and Duarte (2000a, 2009). An additional biographical article including a list of his publications and books citing his works was written by Duarte (2000b). For a first hand account on his early life and travels the reader is referred to his memoirs (Ward, 2004).

4.2 Reminiscences

I think the first time I heard the name John Ward was in the first semester of third year physics. It was my friend, and fellow physics student, Howard Ackland whom using a somewhat mysterious tone referred to a brilliant and famous physicist called John Ward. It was 1976 and we were taking quantum mechanics with Guy Fletcher, a Cambridge physicist, as the lecturer. Fletcher was an excellent teacher he spoke clearly and wrote neatly on the board. For the second part of the course came John Ward. He spoke softly and mumbled a lot. He kept on looking at the board, wrote in a peculiar cryptic style, and kept on shaking coins in his pocket. That was my first impression of the brilliant theoretician. Not a good one, especially since I was interested in getting a good grade in the course.

John was a tall fellow and kept himself in fairly good shape. At the time he was about 52. He wore thick glasses and sometimes stylish clothing. He also taught us solid state physics. As I became more used to his peculiar mannerisms the lectures became clearer and sometimes extremely lucid. This is all I can say about third year physics with John Ward. As I went into my physics honours year, in 1977, John and I began to interact in the physics tea room due to the issue of the B.Sc. Then something extraordinary happened. John Ward, as a full physics professor, had a luxuriously large office. His secretary, Margaret, had an adjoined office. It turns out that for some reason John decided to move out to a small bare office were he took a few books with him. Remarkably enough, he suggested that I and my friend, Milan Brandt, move into his spacious office. So we did. By now the campaign for the B.Sc. had began.

John Ward was somewhat distant and obviously profoundly immersed in his physics. In fact, during this time he even managed to publish a paper (Ward, 1978). I say "managed to publish" because with John, as with many others in the old school of British physics, publication was only considered once a truly new piece of physics was produced. It was quite different to today's anxiety to publish every little extension from an existing work. Besides, in John's case he was highly critical of himself which explains the relatively small number of publications that he authored and co-authored. Some twenty papers, in total. Besides few publications he was a master of succinctness. His doctoral thesis was only 47 pages long (Ward, 1949). His paper on the quantum theory of polarization was less than 2/3 pages long (Pryce and Ward, 1947). The paper that eventually gave rise to the famed *Ward identities* was less than a mere ½ page long (Ward, 1950).

John rarely spoke about his own physics with few notable exceptions. Once going through *The Feynman Lectures on Physics* (Feynman *et al.*, 1965) we came across the quantum representation of the polarization of the photon which was directly related to his 1947 paper and his doctoral thesis (Pryce and Ward, 1947; Ward, 1949). There, in a very tentative and shy manner, he conveyed to us that he was the first to construct those equations. But he did not actually say it in an explicit open manner it had to be almost deduced. I have never seen such modesty in a physicist since. Unfortunately, the ultra brief 1947 note published in *Nature* communicated the main result on the coincidence count rate of two counter-propagating entangled photons but did not include the basic equations describing the quantum polarization states and interference terms. These equations were included in his 1949 thesis but were not published by him in the open literature. More about this matter, later.

The second exception had to do with the application of lasers in isotope separation. Perhaps because of his involvement in the British hydrogen bomb project, or perhaps because of a previous project with the Australian Atomic Energy Commission (Pryor, 1997), he was keenly interested in the performance of narrow-linewidth tunable lasers which was partly the topic of my doctoral research (Duarte, 2010). Following afternoon tea, in an especially dedicated room at the 8th floor of the Mathematics and Physics building, we would come down to the second floor and carry on the conversations in his small bare office. Topics would range from the raging science battles to physics. By this time, he would mostly wear a light brown leather jacket and he would seat on a semi reclining oak chair that made an annoying noise every time he moved. He never bothered to have it fixed.

In retrospect, John's attitude towards the faculty was rather admirable. I never heard him say anything pejorative towards any member of the faculty. In fact he hardly ever said anything about other members of the faculty. He thought highly of teaching and respected those engaged in that activity as he thought that the main function of a university was to provide high quality teaching. In terms of interactions he seemed most at ease with people like Ron Aitchinson, Fredy Chong, Elmer Laisk, Dick Makinson, and later Jim Piper. His own observations on these matters are outlined in his memoirs (Ward, 2004). One matter he did not mention in his memoirs was the disagreement with Peter Mason, the other physics chair at Macquarie at the time. Peter, who was a generalist with interests in biophysics, decided not to support the B.Sc. movement. On that issue John was highly critical of Peter's position but, very wisely, did not make that criticism public.

In general John was very supportive, and regarded highly, the electronic engineering faculty. He had great admiration for anything practical. This admiration also carried over towards experimental physicists. His attitude towards mathematics was different. He expected us physicists to do well in mathematics and I clearly remember him saying "almost all mathematics is trivial." Whether he meant that as a joke, or not, he never elaborated further. However, John perfectly understood the enormous importance of a good mathematical education and he encouraged us to take courses in applied mathematics like Fourier theory, differential equations, and control theory. Being in a unique *School of Mathematics and Physics* a friendly rivalry existed between us and the mathematicians. Probably as a reflection of John's attitude we, the postgraduate physicists, often got into amicable disagreements with some in the math faculty. In one occasion a mathematician, John Corbett I think, responded to our expression of superiority with: "You are just a bunch of plumbers." This was a rather keen observation since we were often talking about getting the vacuum down in our gas lasers. In addition to Professor Chong, the school had several well-known mathematicians including A. G. R. McIntosh, R. H. Street, and of course J. E. Moyal. A perspective on mathematics at Macquarie, during that time, is given by Ann Moyal (2006).

John's benevolent irreverence towards mathematicians, most likely encouraged the attitude of young physicists in the way that we interacted with our mathematical counterparts. Albeit bright, I found them a little bit temperamental and somewhat naïve about practical issues. One day, one of the master students doing research on differential equations was packing his books since he was due to visit a famous university in France. As I observed him, carefully placing his books in a box, I asked: "Are you taking those books, with you?" and he replied: "No, I am shipping them separately." Immediately, I commented: "You'll never see those books again." About six months later he came back, and the first question I asked him was: "Did you get your books?" With an air of sadness, he replied: "No, never saw them again." Another of the math students was very rich and sometimes drove to Macquarie in a beautiful classic Packard automobile. He was also into poetry and was working on a poem for a mysterious classy "silver lady" that he associated with his Packard. For a while, once a week, I would listen to his poem and offer criticisms. He would take notes and come back the following week with a new version of the poem.

Fellow physicists did not escape either. One Friday morning, my friend Milan Brandt, was on the phone while getting increasingly agitated. Suddenly, he handed the phone to me. It was an American who claimed to be very wealthy. He also claimed that his uncle, in America, had arranged for laser beams to be shone on his head. Since we were a laser center, hence his call. My reply to him went something like this: "To steer laser beams from America on to your head here in Sydney, would be extremely difficult to do, have you checked yourself for radiation instead? Then, I proceeded to give him the address where they could have him measured for radiation. Next Monday, Howard Ackland (who was now employed as a radiation physicist), walks into the tea room and asks in a loud voice: "Who was the bloody bastard who sent the crazy American to us?"

It is difficulty to describe John's sense of humor. He did tell a few engineering jokes but what I remember most, is his ability to laugh at situations and events. He laughed particularly hard every time we outmaneuvered or outwitted our political opponents. One hot summer morning, at the tea room, we were informed that some package had arrived at receiving. The postgraduate, for whose experiment the package came in, began to complain at the mere thought of walking to receiving: "Have to walk a whole bloody mile" he said with emphasis in "bloody." At this stage I said: "Not really, you only have to walk 1.61 kilometers." John began to laugh immediately while repeating "only 1.61 kilometers, only 1.61 kilometers."

Around campus, and in physics circles, John was known as eccentric and peripatetic. I guess his incomplete hair cuts, likely done by himself, did nothing to dispel this reputation. Also, I thought it a bit strange that in modern times he still took a boat to America. But what really caught my attention was his house. He lived at 10 Fern Street, Pymble, one of Sydney's most affluent areas, not far from Macquarie. He had a beautiful large house surrounded by tall old trees. The first interesting thing about the house is that it was nearly empty, almost completely devoid of furniture. However, at the center of his spacious living room, with ample windows in the perimeter... a grand piano.

As the population of laser post graduate students grew John decided to teach a course at the honours level called Topics in Physics. This was a course without assigned textbook and he would announce the coming topics as the course went along. Topics discussed included general relativity, further topics in quantum mechanics, and physics relevant to our experimental work. In front of a small group of students, and at this level, he shined as a teacher delivering some very lucid lectures. He would lecture in one class and set up an assignment problem for the next one.

One evening I was in the 3rd year physics lab working on the sealing of a high-purity laser Brewster cell for my molecular laser experiments. A preliminary step in this process was to reach a vacuum in the cell of $\sim 10^{-7}$ torr. So, while waiting, I was on a bench working on one of the quantum problems he had given us. It was about 10 pm as John showed up unannounced. Immediately, his face lighted up as he saw my notes on the problem and cheerfully proceeded to explain some of the subtleties behind operator physics. This was the only time I was given a personal tutorial by John Ward.

At the problem sessions he would call on one of the students to discuss his/her solution to the assigned problem. In one of those sessions I was asked to discuss my solution to a problem related to the diffusion of a charged particle in a plasma within cylindrical coordinates. I proceeded to solve the problem using a Fourier approach. As I went along I noticed John, on my peripheral vision, getting increasingly restless. Unperturbed I continued until I wrote down the final answer. It had taken me about 20 minutes and a lot of equations. John said nothing, he proceeded to clean the board. Half board later, in about five minutes, he arrived at the same solution. Again, the master of succinctness had made his point.

John openly discouraged his physics students from pursuing a theoretical career. The uncertainty of his younger years probably had a lot to do with this attitude (Ward, 2004). As such, he never took a Ph.D. student despite open efforts by some brilliant mathematicians to study under him. More explicitly, John was highly critical of "Ph.D. factories" in theoretical physics. On various occasions he used the word "rackets" to refer to some of these establishments. This issue would resurface later in the 1990s when he was very critical of citations, and the use of citations for promotional ends, despite the fact that he himself was a highly cited author. The point implied by John was that those who directed "Ph.D. factories" were bound to get a large number of citations from their students and former students.

When it came to politics John had no definite ideology. He was a truly independent thinker. However, he cherished individual freedoms and was critical, and even suspicious, of big government. Albeit he thought of democracy as the best system of government he was highly critical of the corruption, and waste to taxpayer's money, sometimes associated with democratic governments. To this end, the thought that the education of the young, via physics and other subjects that required critical thinking, was very important. Also, there was a hidden side of his personality: his humanitarian qualities. In fact, it was John who made it possible for Dick Makinson to come to Macquarie as an Associate Professor knowing full well that Dick had gotten himself in trouble for his left wing tendencies. Also, for a long time, John supported financially the widow of one of his former friends. I learned this by looking at his papers, after his death, as he never uttered a word about it when alive.

In an article written by Greg Sheridan, in *The Bulletin*, John and I were described in the following terms: "One of Ward's few close friends at Macquarie is young Chilean postgraduate student, Frank Duarte, a first class honours physics graduate. The two make an odd couple – the restrained, rather distant Englishman and the intense, earnest South American" (Sheridan, 1980). This was a fair observation. John was born August 1st, 1924. That is almost exactly 30 years and 30 days before I was born. He was born in the Northern hemisphere, I was born in the Southern hemisphere. He was an Anglo-Saxon. I was of Spanish decent who's ancestors had left Europe more than 400 years earlier. And yet, here we both were at Macquarie, in Australia. In my opinion, the factors that cemented our friendship were: physics and an illusive sense of righteousness. An additional common factor was our openness and frankness when it came to expressing a point of view. Then, there were also the peripherals. John admired my political maneuvering and willingness to speak to crowds. And then there was something else: he would often mention my ability to be seen in the company of attractive young ladies. One day that he brought this up, I replied:

"But John, you are *the professor*, why don't you get a beautiful young secretary?... I've seen many in administration." He replied, while laughing in a shy manner: "That's why! *those bastards* get the first pick!"

His directness and frankness often got him at odds with managers and administrators. This was the case at Macquarie where he openly and publicly criticized the status quo and the large education establishment (Mansfield and Hutchinson, 1992; Ward, 2004).

From my perspective, my interaction with John was purely a meeting of the minds, initially. It just happened, either by chance or fate. At first he was the teacher, then the political comrade, and eventually the long time friend.

After I left Macquarie for the US, we maintained sporadic contact via mail. Once he moved to Vancouver, to be relatively close to his sister Mary, he called me and thus began a long series of phone conversations. These were a few long conversations per year and normally they revolved about current issues of physics and my experiences in industrial America.

4.3 The *Arena* interview

Macquarie University had a student newspaper name *Arena*. There, Milan Brant and I wrote a section named Science Notes. In May 1979 we interviewed John Ward (Duarte and Brandt, 1979a, 1979b). Here are a couple of Ward's sentences from that two-part article:

"Einstein did not accept quantum mechanics … he did not accept the statistical nature of the theory…"

"Almost all the serious achievements are simple in principle… the ideas must be sufficiently simple."

In the interview he went on to say that "physics is not a profession but a vocation." In this regard, privately, he had expressed the thought along the lines that "physicists do their best work when near starvation." Perhaps, this thought emanated from his early career.

1979 was the year that we emerged victorious in our campaign to reform Macquarie, and it was also the year that the Nobel for the Standard Model was awarded. Clearly John was annoyed at his exclusion. In this regard, even prior to this episode John would express uneasiness at the propensity for early disclosure by Salam: "My natural instinct was to delay everything until a better, more complete understanding could be reached. Abdus went ahead anyway and in 1958 published a premature effort with me (*nolens volens*) in which he tried to avoid neutral currents. I would certainly have objected to this particular disclosure. This was done right in 1964, but by this time..." (Ward, 2004). Fraser (2008), in his book *Cosmic Anger*, discusses this topic further.

4.4 Quantum Mechanics

In 1981 I accepted a post doc in laser spectroscopy, with Brian Orr, at New South Wales. At this time, I also began to correspond with some US researchers interested in experiments to test Bell's inequalities. When I discussed this with John he said, with his usual frankness: "Don't waste your time." Instead he advised me to work on very sensitive interferometers which, among other applications, would be useful as detectors of gravitational waves. In this regard, like Dirac, he was neither bothered by issues of interpretation nor did he have doubts on the correctness of quantum mechanics. Later, when I had to describe a newly invented *N*-slit laser interferometer I did it using Dirac's quantum notation to the dismay of some of my colleagues. The description is valid for either single-photon illumination or illumination using an ensemble of indistinguishable photons as in the case of narrow-linewidth lasers (Duarte, 1993).

John was pleased with his apprentice. His thoughts on quantum mechanics are beautifully expressed in his memoirs (Ward, 2004):

"The inner mysteries of quantum mechanics require a willingness to extend one's mental processes into a strange world of phantom possibilities, endlessly branching into more and more abstruse chains of coupled logical networks, endlessly extending themselves forward and even backwards in time."

4.5 Epilogue

In November 1999, I got a phone call from John saying that he planned to be in Chile by the end of the year and he asked me whether I was going to be there. As a matter of fact, since my parents had moved back there, I've gone to visit them either in December or January of each year. So, the answer was yes. On December 29, in the morning the phone rang and my brother Henry answered, it was John. He was at the local Sheraton, in Santiago. We agreed on a visit by 11 am. There at the entrance was John. He had flown from Punta Arenas (Southern Chile) where his passenger ship was anchored. Almost 17 years had passed but it was the same John. Older but his usual characteristics intact. Soon, it would also be obvious that his mind was still its brilliant self. We sat down over a bottle of wine as he, with a broad smile, commented that with our shorts and hats we looked Australian more than anything else.

He needed a new pair of glasses and that afternoon we took him to an optics shop, in central Santiago, where his old glasses were measured and a new pair produced. It was a hot afternoon so we stopped at a cheap bar to drink some beer and proceeded to visit an internet café to check on some physics sites. It was then that we noticed that all was not well with John. He became very tired and began to sweat profusely. We took a taxi back to his hotel. At the time, he said he had a cold but that he was feeling better. We agreed
to have dinner next time at our home which was located at the foot of

the Andes Mountains in the East part of Santiago (La Reina). We picked him up from his hotel in my original 1971 Bronco. He clearly enjoyed riding on a truck propelled by a large roaring V8 engine. After re introducing him to my parents we sat on a table under a traditional grape vine canopy. Soon, John and Henry were engaged in an animated and detailed discussion on international finances and Australian tax law which demonstrated yet another facet of John's complexity. During dinner the conversation revolved mainly around geopolitics and the problems Chile faced as a nation including the voracious exploitation of natural resources, by a few, and the adoption of dubious international treaties.

His visit lasted for about six hours. As my brother was about to take him to his hotel he asked me for photographs of my children (he knew two of them as toddlers, Rosa and Frank) and proceeded to quote someone who said that all the money in the world could not equal the value of children. Next, in an unusual personal confession, he expressed regret at not having a family of his own. John had been briefly married in the US and never remarried. He also said that he was really glad to have met my parents again since: "Now I can understand more about you." We met again for most of the day the 1st of January, 2000 at the Sheraton where we had lunch. There, I emphasized to him the need to seek medical attention in Australia on his way back to Canada.

Out of the ongoing conversation, I had with him, there were a number of points worth highlighting. Interestingly enough by now John regarded our *reform of Macquarie* as one of the most important accomplishment of his life. He laughed as he reminisced some of the subtleties of our successful guerrilla campaign.

In the past John had expressed, at least implicitly, annoyance for his marginalization from the Nobel Prize. This time he projected a fairly serene attitude toward this topic and even joked about the existence of more lucrative prizes. However, once again he noticed that another one of his pioneering works (Pryce and Ward, 1947; Ward, 1949) was not being recognized by those publishing in the area. On this topic he was intrigued by a new method of secure interferometric communications, based on Dirac's physics, that I was working on (Duarte, 2002) but we did not have the chance to fully discuss it. He was also interested to hear about an article I was preparing on the prism optics of Newton as related to modern tunable lasers (Duarte, 2000c) and he openly encouraged me to publish it.

An observation is due here: John assumed that other people would automatically acknowledge his work which in today's environment is rather unrealistic. In this regard, he did very little, if nothing, to promote himself or his contributions. He did not write review papers and when, at Macquarie, I asked why he had not written a book he replied: "Because very few people would read it." The lesson here is that even if you do first class physics it is necessary to write reviews and books in order to improve chances acknowledgement.

Beyond physics we discussed topics as diverse as predatory fishing practices in the oceans, the contamination and destruction of the environment in parts of present day Chile (due to large scale mining), and various geopolitical issues. When talking about Marxism he referred to it as an intrinsically flawed ideology. When this came to the surface he briefly reminisced about his friend Dick Makinson, who was a left winger. At one stage he also showed me a manuscript he was working on: his memoirs.

On the 13th of April he called me from Vancouver. He said that due to the bad food in the ship he had almost starved and he added "I have only myself to blame."

He thanked me for his updated page on the internet and laughed when I told him of the small celebration in *The Physicist*, the publication of the Australian Institute of Physics, for the *Ward identities*. I called back on the 20th of April. He recognized my voice immediately. When I asked him how he was doing he replied "I don't know" and after a brief conversation told me to call him back. Some time after this date he was taken to a hospital. John passed away at 12:15 pm on the 6th of May, 2000, accompanied by his sister Mary and immediate family. On the 8th I talked for a long time with Mary. John's mind was clear until the end, she said, and had requested that his memoirs be passed on to me for publication. Finally, he asked to be disconnected from the respiratory machine.

Among the few possessions he had at the time of his death there were some physics papers plus papers and documents related to his memoirs. Also, a few old and nearly disintegrating books mainly on advanced geometry and group theory. To find one of my papers among his depleted set of documents made quite an impression on me.

After John's death, at Mary's suggestion, I got in touch with R. H. Dalitz. Dick and I collaborated for the article published on John Ward by *Physics Today* (Dalitz and Duarte, 2000). Dick was an Australian and a life-long friend of John's. He knew his physics intimately. During an extended series of phone conversations Dick conveyed to me that, following his own careful scrutiny, he had concluded that John's contribution to the Standard Model was more extensive than accepted by popular belief. Incidentally, Dick himself is another giant of post-war physics, known by the *Dalitz Plots* (which are central to the search of the Higgs boson) and *Dalitz Pairs*. Also, recognition of his work has felt short (Aitchison *et al.*, 2006).

During our conversations in his bare office, at Macquarie, John often would go back to his days at Aldersmaston: "*One day I had the idea of radiation implosion. As in all the ideas that have ever popped up in my head, there is no way I can trace the source. I think that one just goes through a mechanical process of trial and error*" (Ward, 2004). The theme came up again during our meeting in Santiago. After his death, while looking at his papers I found correspondence between him and

Lorna Arnold, who was writing a book on the British H-bomb. Although Arnold pays due respect to John's absolute brilliance, as a physicist, she concludes that John's design was an "advanced concept... However, it was not developed" (Arnold, 2001). Dick Dalitz and I spent considerable effort on this subject while we were preparing the article for *Physics Today*. We wrote and discarded many drafts of the article. In one of our discussions Dick said: "We cannot leave it for the historians." The paragraph in our article includes the sentence "Ward had independently conceived a two-stage device, the radiation from the first (fission) stage being used to compress the light elements of the second stage, leading to a thermonuclear explosion " (Dalitz and Duarte, 2000). During the last conversation I had with Dick, after publication of *Memoirs of a Theoretical Physicist* (Ward, 2004), he told me that he was working on a book with a complete collection of John's papers. Note: R. H. Dalitz passed away in 2006.

Following John's death I sent e-mails to a few US physicists who knew him. Philip Anderson wrote back that he had communicated John's departure to Freeman Dyson and then added: "Many thought that he should have been included in the 1979 prize" (Anderson, 2000).

Willis Lamb wrote: "I greatly admired him and his work. I never came close to getting to Sydney, but I would have jumped at a chance to see Ward again. I hope our paths cross again" (Lamb, 2000).

During the year 2000, four articles were published over the life and physics of John Clive Ward: Dalitz and Duarte (2000), Delbourgo (2000), Duarte and Piper (2000), and Duarte (2000b). More recently a book highlighting John's pioneering use of Feynman's diagrams was published by Kaiser (2005) and a description dealing with the Salam-Ward collaboration was published in the book *Cosmic Anger* by Fraser (2008). A summary article based on the material presented here was recently published in *Australian Physics* (Duarte, 2009).

References

Anderson, P. W. (2000). Private communication.

Aitchison, I. J. R., Close, F. E., Gal. A, and Millener D. J (2006). The scientific heritage of Richard Henry Dalitz, FRS (1925-2006). *Nucl. Phys. A* **771**, 8-25.

Arnold, L. (2001). *Britain and the H-Bomb*, Palgrave, London.

Dalitz, R. H., and Duarte, F. J. (2000). John Clive Ward. *Physics Today* **53** (10), 99-100.

Delbourgo, R. (2000). John Clive Ward 1924-2000. *CERN Courier* **40** (6).

Duarte, F. J. (2000a). J. C. Ward: elegant physics and a practical perspective. *Opt. Photon. News* **11** (8), 62-63.

Duarte, F. J. (2000b). John Clive Ward. *http://www.opticsjournal.com/ward.htm*

Duarte, F. J. (2000c). Newton, prisms, and the *opticks* of tunable lasers. *Opt. Photon. News.* **11** (5), 24-28.

Duarte, F. J. (2009). The man behind an identity in quantum electrodynamics. *Australian Physics* **46**(6), 171-175 (2009).

Duarte, F. J. (2010). Atomic vapor laser isotope separation: the Australian contribution, *Australian Physics* **47**(2), 38-40.

Duarte, F. J., and Brandt, M. (1979a). Science Notes. *Arena* **12** (6).

Duarte, F. J., and Brandt, M. (1979b). Science Notes. *Arena* **12** (7).

Duarte, F. J., and Piper, J. A. (2000). John Clive Ward, FRS. *The Physicist* **37** (4), 145-146.

Dunhill, M. (1995). Professor John Clive Ward, in *Postmaster and The Merton Record,* Oxford University, Oxford.

Dyson, F. J. (1949). The S matrix in quantum electrodynamics, *Phys. Rev.* **75**, 1736-1755.

Fraser, G. (2008). *Cosmic Anger*, Oxford University, London.

Greiner W., and Reinhardt, J. (2009). *Quantum Electrodynamics*, 4th Ed., Springer, Berlin.

Kac, M. and Ward, J. C. (1952). A combinatorial solution of the two-dimensional Ising model, *Phys. Rev.* **88**, 1332-1337.

Kaiser, D. (2005). *Drawing Theories Apart: The Dispersion of Feynman Diagrams in Postwar Physics*, University of Chicago, Chicago.

Lamb, W. E. (2000). Private communication.

Luttinger, J. M. and Ward, J. C. (1960). Ground-state energy and many-fermion system, *Phys. Rev.* **118**, 1417-1427.

Mansfield, B., and M. Hutchinson, M. (1992). *Liberality of Opportunity: A History of Macquarie University 1964-1989*, Hale and Iremonger, Sydney.

Montroll, E. W. and Ward, J. C. (1958). Quantum statistics of interacting particles; general theory and some remarks on properties of an electron gas, *Phys. Fluids* **1**, 55-72.

Montroll, E. W., Potts, R. B., and Ward, J. C. (1963). "Correlations and spontaneous magnetization of the two-dimensionaol Ising model," *J. Math. Phys.* 4, 308-322.

Moyal, A. (2006). *Maverick Mathematician.* Australian National University, Canberra.

Pryce, M. H. L., and Ward, J. C. (1947). Angular correlation effects with annihilation radiation. *Nature* **160**, 435.

Pryor, A. W. (1997). Personal memories of two advanced enrichment projects at Lucas Heights in the years 1972-1980. *The Australian & New Zealand Physicist*, **33** (3-4), 53-58.

Salam A. and Ward, J. C. (1961). On a gauge theory of elementary interactions, *Nuovo Cimento* **19**, 166-170.

Salam A. and Ward, J. C. (1964a). Electromagnetic and weak interactions, *Phys. Lett.* **13**, 168-171.

Salam A. and Ward, J. C. (1964b). Gauge theory of elementary interactions, *Phys. Rev.* **136 B**, 763-768.

Sakharov, A. (1990). *Memoirs*, Knopf, New York.

Sheridan, G. (1980). Australian physicist wins Guthrie medal. *The Bulletin* **101** (5239), 49-50.

Ward, J. C. (1949). *Some Properties of the Elementary Particles*, Oxford University, D. Phil. Thesis.

Ward, J. C. (1950). An identity in quantum electrodynamics, *Phys. Rev.* **78**, 182.

Ward, J. C. (1978). General relativity, the Dirac equation, and higher symmetries. *Nat. Acad. Sci.* **75**, 2568.

Ward, J. C. (2004). *Memoirs of a theoretical physicist.* Optics Journal, Rochester.

Ward, J. C. and Wilks, J. (1952). Second sound and the thermo-mechanical effect, *Phil. Mag.* **43**, 48-50.

Chapter 5

Alabama

5.1 The Macquarie papers: tunable lasers

I completed my doctoral research early in 1981 and immediately assumed a postdoctoral position with Brian J. Orr, at the University of New South Wales. Brian was doing double resonance spectroscopy in three atom molecules using a CO_2 laser and a tunable ultraviolet laser. This was quite within my realm of expertise. Whilst at NSW, I had the opportunity to co-supervise a couple of honours students since Brian was partly gone to a sabbatical in Canada. This worked out quite well since it allowed me the evenings to write my thesis and to earn in support of my family. I had gotten married late in 1979, to Patricia, and by the end of 1982 I was the father of Frank Jr. and Rosa.

In December 1981, I submitted my thesis which was entitled *Investigation of the Optically-Pumped Iodine Dimer Laser in the Visible* (Duarte, 1981). This thesis, in accordance to the traditions of the British system was sent to three external referees. Since my research had already yielded two publications, on new narrow-linewidth tunable organic dye lasers, in international journals (Duarte and Piper, 1980; 1981) with more coming, I was fairly confident that it would be accepted. The year was now 1982 and I would usually come to Macquarie in the evenings to either continue experiments on copper-vapor-laser pumped dye lasers or to discuss papers in the works with Jim Piper. At the time, academic jobs were non existent in Australia and I began to look for alternatives especially in the department of defense. There was some precedent for this since a previous researcher at Brian's lab, Gary Nutt, had become a defense analyst.

Thus, I applied for an analyst position at the Ministry of Defense in Canberra and was invited for an interview. I flew to Canberra. They were waiting for me at the airport and I was taken to the Defense Ministry in an official dark Mercedes-Benz, flying small flags, driven by a uniformed driver. There, I was interviewed by a couple of gentlemen and gave a seminar on my recent laser research. When question time came there was not a single physics question. Most of the questions came from high-ranking officers in full uniform and they were all about geopolitics. They were especially interested to know my opinion about the Falkland campaign. I must have made an impression since a couple of weeks later I got a letter with a job offer to join the Australian defense department as an analyst. My job would be to analyze new strategic weapons systems. This was a well compensated and exciting job for life but I also realized that going in that direction would probably mean the end of me as a research physicist.

Interestingly enough, the decision was made for me. It turns out that I had applied for Australian citizenship and my application was accepted. However, somehow, the invitation for the ceremony had been delayed in the bureaucracy and I got it too late. There was nothing that I could do, except apply again and wait in line for another six months or so. Always impatient and with a natural abhorrence towards rituals and ceremonies, I decided not to apply again and thus turned my sight towards America.

One evening in one of my visits to Macquarie I ran into Cathy Foley. Cathy was doing her doctoral research, with Trevor Tansley, in solid state physics and was involved in university committees. It was Cathy, with her usual smile, the first to break the news to me: the referees reports on my thesis were favorable and I should be getting a letter from the university very soon.

My post-doctoral fellowship with Brian Orr did run out at the end of 1982 and by early 1983 I was back at Macquarie doing work on high-average power copper-laser-pumped dye lasers. Jim had managed to secure some extra funding, from the Australian Atomic Energy Commission, and I supplemented my income as a tutor at the University of Sydney and the New South Wales Institute of Technology. It was during this uncertain and exciting time that the multiple-prism grating dispersion theory was introduced (Duarte and Piper, 1982; 1983), new tunable oscillator cavities were reported (Duarte and Piper, 1982b), and copper-laser pumped tunable multiple-prism grating oscillators, operating at 8-13 kHz, were demonstrated (Duarte and Piper, 1984a; 1984b).

One afternoon in May 1983 I graduated for the first time since I had refused to graduate back in 1978 in protest against the degree system. It was a wonderful sunny afternoon and I dressed to the occasion wearing from a pair of fine German leather shoes to a fine silk crimson bow tie, courtesy of Patricia. Following the graduation ceremony, in a garden reception at Macquarie, a large number of friends from the sciences came to say hello. Both, Jim Piper and John Ward seemed to be sincerely content. Frank Jr. and Rosa were too young to remember any of this. That evening the whole family went to one of the finest restaurants in downtown Sydney where we dined in a reserved room with lobster as the main course. I remember this as one very special day in my life.

During this time, besides my academic work, I also designed a laser perimeter alarm system and even applied for an Australian patent. My father, Luis, soon built a prototype and designed other types of alarm systems that he kept on perfecting and selling to customers, again by word of mouth.

By now, I had applied to a position involving laser research at the University of Alabama in Tuscaloosa. Events followed in rapid succession and one evening in August, I found myself in a farewell walk on Macquarie's grounds. I was leaving the next day and I just wanted to take a final look at the Macquarie I knew for nearly 10 happy years. As I was leaving, a friend waved and we sat on the grass talking for a while. "I wish you the best in America" she said and game me a soft kiss. This was my good bye from Macquarie.

5.2 Alabama: high-power tunable CO_2 lasers

One afternoon in early August 1983 I left for Sydney's international airport in the company of my family and two physics friends: Haruki Tsuda and Geoffrey MacKellar. Haruki was doing his masters in miniature metal vapor lasers and Geoff was doing his doctoral research in high-power gas lasers. I left alone. Patricia and the kids would stay behind until later. I boarded a crowded DC-10 to Toronto. There, I went through US Customs. Then, on to Atlanta. At Atlanta's airport I could not readily find the terminal for my connection. After some asking I ended at the very end of one of the secondary terminals where some men were working on a small two engine high-wing plane. A couple of hours later they called me and the other passenger to come on board. The pilots had to come back to Atlanta so they did not bother to gain much altitude and at about 1500 ft altitude we made it all the way to Tuscaloosa. There, the return-flight passengers were waiting and in the group a smiling face. He was Philip Coulter the Chairman of the physics department.

Although a native Southerner, Phil was a theoretician educated in California. That evening he took me to his home where his wife, in a traditional expression of Southern hospitality, cooked some delicious fried chicken plus an assortments of salads. After dinner, we talked for a couple of hours and Phil said that he would get me some campus accommodation next day. After breakfast, and a brief introduction to the department, and my new office. Phil took me to administration to secure a room. There, I found myself in the hall of an old building (probably Civil War vintage) being used now as office space. There were about half-dozen blond women all talking with their distinct Southern accent. For a moment I thought I was in some sort of movie set but quickly came back to reality when Phil, who spoke mainstream American English, handed me a set of keys for my new room.

The campus was nearly empty and it was hot, indeed, bloody hot and humid. It was so bad that I tended to stay in my air conditioned office until 2 or 3 in the morning working on my lecture notes. In all honesty, for a few times I thought I had made a big mistake coming to Alabama. That was my introduction to the United States.

Soon I realized some odd features. The senior professor who hired me was also relatively new to Alabama. Strangely enough, he was no where to be found. It turns out that during summer time, he would go back to Quebec and his assistant would look after his large house, near a lake, and he would not come to the lab either. So, no laser research got done in summer time.

Their field of research was far infrared (FIR) sources. When the senior professor came back I quickly realized that he did not drink tea or coffee so there was little room for interaction. Thus, we drifted apart before we had any chance to work together. On the other hand, I interacted quite a lot with other members of the faculty among them Phil Coulter, Jack Sulentic, Dick Tipping, and Pieter Visscher. Phil tried to make me feel at home and he often invited me to have lunch with the dean and also took me to a couple of Alabama football games. Both, Coulter and Visscher, were familiar with the *Ward Identities*. Also in the faculty there was an old timer called James H. Bartlett. He had come from the University of Illinois and was known for his pioneering contributions to atomic physics. A grand old gentleman of physics him and I would talk from time to time, in the little library of the department, about physics and other topics.

My teaching went very well. At the end of my first semester the whole engineering class applauded and I was given very good evaluations. By the end of the second semester student from other classes were coming to sit in my lectures. The best outcome of this exercise was that I inspired a selected few, who originally had come for engineering training, to switch to physics.

The optics laboratory that I came to work in had the lasers installed high, just below eye level (this is not a misprint) on a rail type system. In this particular lab the lasers were installed around a massive central remnant of a previous NMR system. This was the elephant in the room. The only thing usable to us was a diffusion pump which I salvaged. Early in the summer of 1984 I got tired of the massive structure and went to Phil to ask for permission to remove it.

Perhaps thinking that the apparatus was too massive to be removed Phil said with a smile: "Sure, go ahead." Phil did not know me well enough nor did he know that I had two mechanically apt students working with me. We started after lunch time and by 3 pm a small crowd had gathered to watch how we were taking the monster apart piece by piece. It was a feat of cleverly applied strength. When Phil showed up about 2/3 of the job was done and it was by now irreversible.

Always impatient and quick to action, I began my own little research program. I continued work on the dispersive theory of multiple-prism expanders (Duarte, 1985a) and became the first to introduce multiple-prism grating configurations for linewidth narrowing in high-power tunable CO_2 lasers (Duarte, 1985b, 1985c). The prisms were fabricated out of ZnSe. I also published several papers in conference proceedings.

I spent the rest of the summer working on my CO_2 laser experiments now in a far more spacious lab. Late in 1984, the senior professor paid a rare visit to my office to complain about my aloofness from the FIR work and proceeded to attempt to exert some pressure, using my coming tenure review as leverage, in an apparent effort to steer me towards FIR research. This type of tactics provoked a reaction in no uncertain terms from my part. I brought his rare visit to a sudden end and he left my office in a clear hurry. In turn, this initiated a political confrontation in the department. Since I had come all the way from the Southern hemisphere to the US to do physics, and not politics, I decided to resign. Following my decision, several of the professors quietly expressed their support. One day I met with Dr. Bartlett who gave me a copy of his book on classical mechanics (Bartlett, 1975) with a brief dedication and signature. Also, during my stay in Alabama I made good ties with the postgraduate students (Robert Bass and Wayne Keen, in particular). They, like me, would sometimes work until the early hours of the morning.

The job search was a little sterile for a while until I got a helpful hint from one of the secretaries, Corinne Ray. It turns out that one of the local references spoke highly of my physics but at the same time said that I was occasionally difficult do dealt with. As soon as I replaced that reference the search progressed and early in 1985 I had a firm job offer from a Southern university, at the rank of associate professor. About the same time I was invited to give a talk to the prestigious Photographic Research Laboratories of Eastman Kodak, in Rochester, New York. The same afternoon, after my talk, I was made a job offer which I initially did not take seriously. Following repeated phone calls, the offer of a substantial salary, plus a generous signing bonus, I accepted. Thus, I was hired at the rank of Senior Research Physicist at one of America's top research institutions. Part of the bonus was used in buying a new Pontiac Parisian that was used to drive up north to Rochester, New York, with Patricia and the kids. At this stage, my plan was to spend no more than five years in industry.

My stay in Alabama can be summarized as two very creative, fun, exciting, and sometimes tumultuous years. The turbulence and chaos included marriage difficulties, and a temporary expulsion from home, given an affair with a young Southern lady.

References

Bartlett, J. H. (1975). *Classical and Modern Mechanics*. University, Alabama.

Duarte, F. J. (1985a). Note on achromatic multiple-prism beam expanders. *Opt. Commun.* **53**, 259-262.

Duarte, F. J. (1985b). Variable linewidth high-power TEA CO_2 laser. *Appl. Opt.* **24**, 34-37.

Duarte, F. J. (1985c). Multiple-prism Littrow and grazing-incidence Pulsed CO_2 lasers. *Appl. Opt.* **24**, 1244-1245.

Duarte, F. J. (1981b). *Investigation of the Optically-Pumped Iodine Dimer Laser in the Visible.* Macquarie University, Sydney.

Duarte, F. J., and Piper, J. A. (1980). A double-prism beam expander for pulsed dye lasers. *Opt. Commun.* **35**, 100-104.

Duarte, F. J., and Piper, J. A. (1981). Prism preexpanded grazing-incidence grating cavity for pulsed dye lasers. *Appl. Opt.* **20**, 2113-2116.

Duarte, F. J., and Piper, J. A. (1982a). Dispersion theory of multiple-prism beam expanders for pulsed dye lasers. *Opt. Commun.* **43**, 303-307.

Duarte, F. J., and Piper, J. A. (1982b). Comparison of prism-expander and grazing-incidence grating cavities for copper laser pumped dye lasers. *Appl. Opt.* **21**, 2782-2786.

Duarte, F. J., and Piper, J. A. (1983). Generalized prism dispersion theory. *Am. J. Phys.* **31**, 1132-1134.

Duarte, F. J., and Piper, J. A. (1984a). Narrow-linewidth, high-prf copper laser-pumped dye-laser oscillators. *Appl. Opt.* **23**, 1391-1394.

Duarte, F. J., and Piper, J. A. (1984b). Multi-pass dispersion theory of prismatic pulsed dye lasers. *Optica Acta* **31**, 331-335.

Chapter 6

Lasers and the Cold War

6.1 US Army Missile Command

For a while I had been attending an old conference called *The International Conference on Lasers*. In December 1984, I went to San Francisco. I had already resigned from Alabama and had not found a job yet. I was also running into financial difficulties so I decided to save money by staying at a cheap hotel in downtown San Francisco. This was an old wooden building and my room was in the second floor. A glass in the window was broken and there were strange noises coming from the other rooms. Thus, I did not unpack and that night rested over the bed with my clothes on. Next morning, a Sunday, I left the hotel very early and began walking towards the conference hotel while inquiring in a few places for "cheap" accommodation. Following an unforeseen incident while trying to book a room at another cheap hotel, I ended up at the expensive place I was supposed to be, anyhow. At least I saved a little the first night, I told myself.

At the conference, I gave a couple of papers. I must have made some kind of impression because C. P. Wang, from the Aerospace Corporation, invited me to be part of the program committee for *Lasers '85*. Unknown to me, Charlie had also invited to his committee a senior US Army physicist called R. W. Conrad. One morning, early in 1985, I was at my office at the University of Alabama when the phone rang. A gentleman with a very distinctive voice, with a slight stuttering, introduced himself as "Ray Conrad, US Army." We talked a little about the conference and then he asked me what I knew about high-power dye lasers. I proceeded to explain my expertise in the field. Towards the end of our conversation he invited me for lunch at Huntsville, Alabama, home of Redstone Arsenal and the US Army Missile Command.

The day of the visit, after going through security, I arrived at his building late morning. Following a quick tour of his lab we headed to a steak place to have lunch. At first the conversation focused on high-power dye lasers then, as the beers went by, our conversation drifted into geopolitics and the Cold War. It was a long lunch and Ray and I became instant friends. Up on departing, he said that he would look into doing some joint experiments in dye lasers. That was the beginning of a 13 year interaction between me and the US Army Missile Command (MICOM). After that, the association continued with the newer US Army Aviation Missile Command.

When I decided to go to Eastman's I told them that I would like to continue my interaction with the US Army. My division director, Joe Merrigan, told me that it was fine. To this effect I arranged to visit Redstone Arsenal during summers using my vacation time.

Since my army colleagues were interested in high energy pulses, I began doing work on flashlamp-pumped dye lasers. According to the Soviet literature these lasers could deliver up to 400 J per pulses (Baltakov *et al.*, 1974). That is an enormous energy unfathomable for even trained physicists. As an example I should say that most pulse lasers used in laboratory research are at the milli Joule (mJ) level. That is, at one-thousandth of one Joule. More about the Soviets later.

Ray and I had no research plan, no set budget, no time limits. Also, we had no bureaucracy over us, Ray had taken care of that. If Ray had a plan he did not convey it to me and I did not ask. It was as close to an ideal open ended research situation as can exist. As far as I knew, the aim was to do good *tunable laser research* eventually producing highly coherent lasers at high energies. The word tunable here is critical. These are lasers whose wavelength, or color, can be varied at will (Duarte, 2003).

Thus began a series of experiments. I gave Ray a list of optics parts, a list of optomechanical parts, and they were to have the flashlamp lasers working during my visits. Partly because of my experience with very narrow-linewidth laser oscillators at Macquarie we set out to apply the same multiple-prism architectures to long-pulse dye lasers. The oscillator is the "brains" of any laser system so it was an important thing to do. Thus, successful experiments were performed and subsequent papers published (Duarte and Conrad, 1986, 1987). At around 1988 Ray was moved to a secretive project and I continued work with J. J. Ehrlich and some younger physicist notably W. E. Davenport and T. S. Taylor.

Ray was a real character. He was quite tall, about 6 inches taller than I, and he was always listening to a small radio. It was the weather. He lived in his farm, near a village called Russellville, about 70 miles West of Huntsville. When the weather was bad he drove a big Ford Ltd. When the weather was good he flew his plane from his farm to an airport near by and back. Since the Alabama weather in summer time can change for the worse quite rapidly, he needed to be hooked to that little radio. One Sunday morning I drove to his farm, in a Chevrolet Impala, on a wide empty road. Quickly I gained some understanding on the logic behind Ray's flying. I still remember the pleasure of pushing that V8 to its limit while thinking how much better flying would be. In his farm he lived with his wife, a college professor, had a small vineyard and a landing strip. He also had a pond where he did his fishing. One day in the late 1990s Ray called me to tell me that he was taking an early retirement while he was still young and healthy. He would still do physics but at his own pace. His decision did not surprise me.

Slowly we began to tune our sophistication in planning and performing experiments. The younger physicists were quite efficient and disciplined and thus we kept publishing new findings on narrow-linewidth long-pulsed tunable lasers (Duarte *et al.*, 1988, 1990, 1991, 1997, 1998).

Among the accomplishments of this flashlamp-pumped tunable dye lasers program were: the first demonstration of long-pulse narrow-linewidth multiple-prism grating oscillators (Duarte and Conrad, 1986); first demonstration of oscillator-amplifier configurations with this class of oscillators (Duarte and Conrad, 1987); first study of linewidth instabilities in this class of oscillators (Duarte *et al.*, 1988); first demonstration of a long-pulse narrow-linewidth solid-state dye laser oscillator (Duarte *et al.*, 1998).

The pride of this research effort was our ultra-stable narrow-linewidth multiple-prism tunable oscillator described by Duarte *et al.* in 1991. As far as the open literature goes this is the first, and only, *ruggedized* narrow-linewidth dye laser oscillator built to date. Besides incorporating an optimized dispersive architecture it was engineer to last in an *all-super-invar* structure. As far as the sealed dye system goes, all the tubing was stainless steel and teflon. This ruggedized oscillator yielded single-transverse-mode, single-longitudinal-mode emission. In English that means extremely pure tunable laser emission. This oscillator was displaced on a vehicle over a rugged terrain with no adverse effects on its alignment or laser performance.

I was never told, by Ray or anyone else, why the US Army was interested in high-power tunable lasers. And again, I did not ask. However, it is not all that difficult to come up with a plausible scenario. First, when Ray called me in 1985 we were at the height of the Cold War and it was also known, even from the open literature, that the Soviets had a healthy interest in this field of research. Thus, it would not be unreasonable to expect the US counterparts to have their own interest. Second, it is obvious that in a modern tactical theater access to wavelength agility, that is tunability, is essential. This is also the case if you are trying to anticipate and to create defenses against tunable laser radiation. By the early 1990s, with the end of the Cold War, the funding for this type of research came to an end. At the time this happened, other laser colleagues in the US had achieved substantial success in high-energy high-average-power amplifiers. Also our British colleagues had developed their own successful program.

This is how things would work: we were almost continuously engaged in brief conversations about high-power tunable lasers. This involved comments on newly published research and our own new ideas. When we thought that we had something new, and useful, then my colleagues at MICOM would start dealing with all the necessary bureaucratic details and to order what was needed. When everything was in place I would show up, often during summers using my vacation time, for a couple of weeks to do the experiments. This not always went smoothly and sometimes working around the clock was necessary.

In every occasion the final days of my visit would revolve at tiding up details and organizing the data collected. During these days we would have some fun after work. One such occasion involved a visit to a local bar. We were happy and even a little bit euphoric about the success of our ruggedized oscillator, thus the beer began flowing rather rapidly. At one stage, I noticed the eyes of one of the waitresses serving our table. They were of a beautiful deep green color. Surely enough I came up with an incredible line: "Your eyes are beautiful… they are as green as coumarin 504." At first, she looked at me bewildered as if I was speaking in a foreign tongue but displaying her Southern politeness she replied with a gracious "Thank you." As soon as she left Wayne and Travis began laughing. "I bet that's the worse line she has ever heard… coumarin 504!" said one of them. Anyhow, as bad as it was it was not a disaster since she had no trouble coming back later with a group of her colleagues to participate in a little ceremony to induct me as "weenie man of the hour." This involved me drinking beer standing on a table while they went around the table chanting something incomprehensible in their high pitch voices.

By the early 1990s, the success of our tunable laser program was noticed by one of the top army physicist at The Pentagon and John Ehrlich and I were invited to give a talk. As the date of the visit was fast approaching the issue of me not yet being a US citizen surfaced.

Then I was told to approach "my government" to get something like a letter of introduction according to diplomatic protocol. To this effect, the top brass at MICOM sent a letter to the representatives of my government in Washington DC. The letter highlighted the importance of my work to MICOM, etc. Needless to say there was never a response: neither a yes nor a no. Then, as I was contemplating canceling the trip I got a call from our host at The Pentagon to "show up anyway." John and I converged on a Sunday, had dinner together, and went over our presentation. Still we did not know how the "paper issue" was going to be resolved. All we knew was to show up at 8 am sharp at one designated entrance. At the exact time, a security escort showed up. They asked for our names and identification and immediately I began walking, at the center of the group, to our destination. Later, as we were in a coffee break, I overheard a secretary complaining about the eccentric scientists they had to deal with "some of whom required special attention." Apparently, for physicists, no bureaucracy is insurmountable.

In the early 1990s, the attention of the US Army focused on solid-state dye lasers. In 1992 we reported, at a conference, for the first time on a dispersive solid-state dye laser oscillator (Duarte *et al.*, 1993). This was the ground work that eventually led to an elegant multiple-prism grating solid-state dye laser oscillator (Duarte, 1994). The 1994 paper represented a breakthrough and became widely cited in the field. Further work in this area led to Duarte (1997, 1999) and Duarte *et al.* (1998). By 1998 the ruggedized narrow-linewidth oscillator built in 1991 was still working flawlessly.

6.2 The laser conferences

In the late 1980s, via Jim Piper, I became acquainted with a laser conference called *The International Conference on Lasers 'XX* where the *XX* were substituted by the last two digits of the current year.

Usually, these conferences took place during the first week of December. I first went to this conference in 1979 when it was held in Florida. The conference at the time was one of the most popular conferences of the laser community with particular high attendance of laser physicists and engineers working on high-power discharge lasers. Additional attractive aspects included a well organized hard bound book series entitled *Proceedings of The International Conference on Lasers* 'XX. Further positive aspects included the lasers mainly nature of the conference which enabled interaction with well known laser physicist at relatively amicable surroundings in absence of the large crowds associated with many conferences today. The attendance of a few hundreds enabled organization of the conference at attractive locations such as New Orleans, Las Vegas, and Lake Tahoe. Unusual aspects of the conference were that it was not associated with the usual large professional bodies and was directed by The Society of Optical and Quantum Electronics. The Program Chairman would be selected from the laser community every year and the Conference Chairman would always be V. J. Corcoran. Vince Corcoran was an odd character who had done some good research in non-linear optics in the 1970s. He owned prime real state in Alexandria, Virginia, where he had a laser company. Apparently, he was a very good businessman but his personality was not the friendly type. However, he did have a great asset in his favor: once you were selected Program Chairman he gave complete and total organizational freedom.

I was first brought to the Program Committee in 1985 by C. P. Wang from the Aerospace Corporation. That year the conference was organized in Las Vegas and Charlie had the idea of organizing a panel discussion session on Ronald Reagan's Strategic Defense Initiative (SDI). Participating in the panel were:

Charles A. Brau (Los Alamos National Laboratory)
James A. Ionson (SDI Office)
Lois C. Marquet (SDI Office)

C. Kumer N. Patel (Bell Laboratories)
Saul S. Penner (University of California at San Diego)
Charles K. Rhodes (University of Illinois at Chicago)
Robert Roberts (Institute for Defense Analyses)
Robert C. Sepucha (Space Technology)
Edward Teller (White House Science Council)
Gerold Yonas (Chief Scientist at SDI Office)

For a transcript of this panel discussion the reader should refer to Wang (1986). The keynote speaker was C. K. N. Patel, inventor of the CO_2 laser. This conference was a notable success and the Program Committee even received a congratulations telegram from President Reagan.

In 1986 the conference was held in Orlando, Florida. This year the conference was chaired by R. W. McMillan from the Georgia Institute of Technology and again I organized a session on dye lasers. This year attendance was down, relative to 1985, but my session was very active and well attended. Perhaps this was the reason that Bob McMillan suggested that I should chair *Lasers '87*. When Bob communicated his suggestion to me, I responded with cautious optimism but asked him for some time to think about it. I then had a conversation with Ray Conrad who was somewhat skeptical about the benefits of such venture. Defending my optimism I recall telling him: "Ray, this could really be an experience for me." To what Ray replied: "Breaking a leg is also an experience." Next day, Bob McMillan approached me again to ask if I had made a decision. "Yes, I will do it" was my answer. Once I told Ray of my decision he congratulated me and promised to help in any way he could.

6.2.1 *Lasers '87*

My strategy for organizing *Lasers '87* was simple: it consisted in building on the existing program committee to create an extended and reinforced committee comprising more than 60 members many of whom came from well-known institutions. Next, I made a list of the best-known laser scientists at the time and proceeded to invite them to deliver plenary addresses. I also decided to bring back the panel discussion on SDI including well-known members of the community. Next, I began to work on raising funds to pay for expenses of deserving speakers. This included conference registration fees and assistance toward travel expenses for some. To this effect, I approached well-known corporations and government agencies. The list of plenary speakers included:

C. A. Brau, Los Alamos National Laboratory
R. W. P. Drever, Caltech
K. Fujii, Ibaraki University
L. W. Hillman, Cornell University
R. T. Menzies, Jet Propulsion Laboratories
E. Moses, Lawrence Livermore national Laboratories
J. Miller, TRW
M. O. Scully, Max-Plank Institut
W. T. Silfvast, Bell Laboratories
C. R. Stroud, University of Rochester
C. R. Vidal, Max-Plank Institut
C. E. Webb, University of Oxford
L. E. Wilson, US Air Force Weapons Laboratory

The panel discussion on SDI was composed of:

R. G. Briggs, Lawrence Livermore National Laboratories
L. Goldman, University of Cincinnati
J. D. Hammond, SDI Office
J. A. Ionson, SDI Office
T. H. Johnson, US Military Academy
A. R. Kantrowitz, Dartmouth College
C. K. N. Patel, Bell Laboratories
W. Seka, University of Rochester
R. Sproull, University of Rochester

E. Teller, Hoover Institute

In addition to the list of plenary speakers there was a strong presence of Soviet physicists who delivered an ample number of papers including well-known authors such as V. V. Apollonov, V. Y. Baranov, N. G. Basov, and A. M. Prokhorov. In addition to the Soviet delegation there was also a delegation from Portugal. The success of *Lasers '87* was summarized in the largest attendance of this series of conferences which was around one thousand participants. For a narrow-focus specialized conference that was quite an achievement at the time. The book containing the conference proceedings exceeded 1100 pages (Duarte, 1988) and became the largest in the series.

All of my dealings while organizing *Lasers '87* were not done with Vince Corcoran but with his capable and gracious daughter Kathy, who took over the logistics of the conference at the time. Kathy and I got along fairly well and I had her full cooperation and support. In addition, I had a blank check from Eastman Kodak to spend time in the organization of the conference. A further factor in the success of *Lasers '87* was the selection of Lake Tahoe as the venue.

There were so many good papers submitted to *Lasers '87* that I had even to organize an evening session on nuclear-pumped lasers which I ended up chairing. This was on the Monday and the session was very well attended. Early in the session I saw Lloyd Hillman making faces and gestures at the door of the conference room. Just to silence him I went to the door and asked him what he wanted. "The key to your room" Lloyd responded. Without thinking twice I gave him the key and went back to the session. As it turns out, the only material thing I got out of organizing this conference was a presidential suite at the place of the conference which was the Harvey's Resort Hotel. A couple of hours later when the session was over, I headed back to my room. Still thinking about the session I had even forgotten that I had given my key to Lloyd. As I stood in front of the door searching for my key I heard voices and noticed that the door was not fully closed. As I opened the door I found that the room was full of people… and bottles of wine every where. There were also generous trays with cheese, bread, grapes, etc. I recognized about half the people. I had never seen the other half but Lloyd assured me they were "All from the conference." The party went on until the early hours. Next day, I woke up surrounded by empty bottles of wine. Still, with a slight headache I went to chair the morning plenary session. At mid day I headed back to the room and in the corridor I overhead a couple of maids talking in a scandalous tone about one of the suits that was found "full of empty wine bottles ." (Note: Lloyd Hillman died unexpectedly in 2005).

In organizing the panel session on SDI, I tried hard to get a balanced composition of the panel. To that effect I invited, in additions to those who attended, N. Bloembergen and Hans Bethe. Both of them declined to attend but Hans Bethe was quite helpful in suggesting alternatives critical to the SDI concept. In addition, I invited R. P. Feynman whom, via correspondence, valued the invitation but declined to attend due to his already fragile health.

Edward Teller arrived late morning on Wednesday (December 9, 1987) for the panel discussion held that afternoon. His entrance to the main conference auditorium, surrounded by a praetorian guard, was quite a sight and something that I have never seen in a physics setting again. At the panel discussion, there was standing room only. Salient remarks from Teller include: "It is wrong to postulate that SDI won't work unless it works completely" and "The Soviets have not only gone ahead with SDI, as Gorbachev has now stated, but we also know that they have vastly outspend us on lasers." He also criticized the aspects of classified research with statements such as "Secrecy in science does not work." In addition, he pronounced a mild criticism toward the American Physical Society (APS) report on SDI and engaged in a polite debate with Dr. Patel who was one of the authors of the report.

An interesting question posed from the floor to Dr. Patel was: "Dr. Patel stated here, and also to *Physics Today*, that the APS report nowhere says that SDI won't work." To which Dr. Patel replied: "That is correct." For the complete version of the panel discussion, including questions and answers, the reader is encouraged to read the proceedings of the conference (Duarte, 1988).

Personally, what I found most instructive in this exercise was Arthur Kantrowitz's discussion of laser propulsion. Following the panel discussion there was a press conference, focused on Edward Teller, where questions on the Livermore X-ray laser were asked. Subsequently, Teller game me a copy of a paper he had co-authored, on superconductivity, and which eventually I ended up editing for publication in the proceedings (More and Teller, 1988).

These were the SDI years and the golden age in high-power laser development with large projects in excimer lasers, metal vapor lasers, dye lasers, free electron lasers, and chemical lasers. One evening, at one of the Harvey's bars I asked Ray Conrad if he would have money to finance some new experiments. Ray answered in his typical and slightly stuttering style: "Frank... I have money... to spray people with."

6.2.2 *Lasers '88* and *Lasers '91*

Lasers '88 also took place in Lake Tahoe. This time I was co-Chair with Bob Sze from Los Alamos. This conference was low key relative to *Lasers '87* and my participation in its organization was almost on automatic pilot. The key note speaker was Charlie Brau who spoke on the development of very high-power lasers. Plenary speakers included

E. V. Browell, NASA
J. C. Diels, University of New Mexico
R. L. Fork, Bell Laboratories

S. Haroche, L'Ecole Normale Superiore
L. W. Hollberg, NIST
F. K. Kneubuhl, Swiss Federal Institute of Technology
B. Newnam, Los Alamos National Laboratory
J. Schwinger, UCLA
H. Walther, Max-Planck Institut

There was also a panel discussion on high-power dye lasers which included:

H. R. Aldag, AVCO Research Laboratory
R. W. Conrad, US Army
F. J. Duarte, Eastman Kodak Company
P. N. Everett, MIT Lincoln Laboratory
J. A. Paisner, Lawrence Livermore National Laboratory
T. G. Pavlopoulos, Naval Ocean System Center
C. R. Tallman, Los Alamos National laboratory

This panel discussion was very technical for the most part. However, the most interesting input came from Ray Conrad who after listing some extraordinary recent accomplishments, such as the demonstration of ~ 800 J per pulse in an excimer-laser pumped dye laser and the attainment of ultra narrow-linewidh emission (Duarte and Conrad, 1987), posed a very relevant question. The question related to the future and: "Who is actually going to fund open-ended questions like these?" He then proceeded to explain that he saw DOD investments in these areas going down. This was the first indication that the golden age of high-power laser development was about to come to an end. At least in that cycle. Again, the complete panel discussion is included in Sze and Duarte (1989).

On Thursday morning, following Walther and Haroche, it was Schwinger's turn for his plenary. The Chairman of the session was Marlan Scully. Schwinger began to deliver a paper on coherence and the Stern-Gerlach experiment. At the end of his 30 minutes it soon became obvious that he was not going to stop. Five minutes, ten minutes, then I said to Marlan: "He is going to ruin the timing for all remaining sessions, please tell him to stop." Marlan replied: "But Frank… he is Schwinger!" Scully's attitude reflected the deep respect that Schwinger inspired in the physics community where he was referred to as a "prince of theoretical physics" (Kaiser, 2005). Realizing that Marlan was to continue his silence I then approached Schwinger, taped him softly over the shoulder, and said with an air of benevolent authority: "Professor… you are ten minutes over, please stop." Immediately Schwinger replied: "Very well."

That evening there was a reception in which Scully formally introduced me to Schwinger. As a conversation developed, between the two of us, I commented that the hardest thing about the Stern-Gerlach experiment was to reach the necessary ultra low vacuum. Schwinger: "Where did you learnt about Stern-Gerlach?" I replied: "At Macquaric." Schwinger: "That's where John Ward is." To which I added: "Yes, he was my professor." Then, Schwinger smiled broadly and proceeded to ask: "Is he difficult to get along with?"

Next morning I was having breakfast with Marlan Scully when Kathy Corcoran spotted us and came almost running to hand me a piece of paper after saying "This is from your director at Eastman's and is very urgent." I concluded breakfast with Marlan and headed to my room trying to guess what in the world would have caused a director to call me. Sure enough, the phone number was from the Director of the Photographic Research Laboratories. That was the boss of my boss of my boss. He proceeded to tell me that he had received a call from Colby Chandler, the CEO at the time. He then explained that Chandler was bit concerned about some "Russian visitors." I clarified that they were visiting the city of Rochester and not my laboratory. Next, he told me that some government people would visit me and asked me to be "nice." Next Monday, once at Rochester, I called a friend and explained the situation. He said he would call me back within an hour or so. The phone rang and my friend explained that the people that wanted to talk to me were the FBI and one of the intelligence agencies. He went on to say: "The FBI is concerned with espionage and the intelligence people would probably ask you to help them." So I relaxed.

Within a couple of days two gentlemen called to set up appointments. Just as my friend had said, the FBI asked a whole bunch of questions trying to ascertain who these Russians scientists were and what they wanted. Again, as he suggested the intelligence people took a different approach and after a long conversation they proceeded to ask for my help. In the same conversation I replied that I felt uncomfortable about that type of activity and I politely declined the offer. Nevertheless, I also said that I would have no impediment to help with technical matters. For the next few years, about twice a year, I was visited by technical analysts who often had a bunch of questions related to new laser and optical technology trends.

Back to late 1988: as it turned out Apollonov and a couple of his colleagues did show up in Rochester and spent a couple of days visiting the local universities and talking to local academics. I did not invite them to my lab albeit I took them out to dinner to a local Mexican restaurant. At dinner it became apparent that even at this late stage, members of the Soviet elite still held on to Marxist ideas.

Participation of high-ranking Soviet scientists continued to increase in *Lasers '89* and *Lasers '90*. Again I was co-Chairman of *Lasers '91*. More of the same, with one twist, the Soviet Empire was near collapse and coming to an end. At the same time, the participation of important Soviet scientists in the conference had proportionally increased. By now, it was apparent that the end of the Cold War was on thus resulting in the beginning of a drain in funding for high-power lasers and directed energy research. Of particular relevance to this conference was a panel discussion on high-power lasers with the following participants:

A. Budgor, US Navy
N. Cheburkin, Astrophyzika, USSR
V. Danilychev, Innovative Laser Systems
N. Griff, SDI Organization
L. E. Wilson, Rockwell International

Without doubt the significant aspect of this discussion was the presence of Nikdi Cheburkin supreme head of Astrophyzika. This was an enormous defense establishment whose existence in the West remained unknown until the end of the Cold War. The business of this organization was the design and manufacture of high-power lasers for military applications. For some unclear reasons the bulk of the remarks of Cheburkin were not available at the time of publication of the proceedings. The Cheburkin remarks did refer to a large MW class CO_2 laser capable of correcting the orbit of satellites. The rest of the discussion was published in the proceedings (Duarte and Harris, 1992).

After *Lasers '91*, I continued to participate in the Program Committee and more and more I began working with Marlan Scully in the awarding of a medal called the *Einstein Prize for Laser Science* whose recipients included R. Chiao, J. L. Hall, S. E. Harris, T. W. Hänsch, H. J. Kimble, W. E. Lamb, L. M. Narducci, N. F. Ramsey, W. F. Walls, H. Walther, C. E. Wieman, and D. J. Wineland. By 1995 this list had two existing Nobel laureates and three future Nobel laureates.

At *Lasers '92*, held in Houston, I had the opportunity of having some long conversations with Nobel laureate W. E. Lamb. Willis Lamb is famous for the *Lamb shift* and the *Lamb dip*. Among theorists, Lamb is referred to as a "rare theorist turned experimentalist" (Kaiser, 2005). I had recently reviewed a book on Dirac (Duarte, 1991) and I was also very familiar with some papers on the measurement problem. These conversations also included the topic of John Ward. Lamb knew John personally and he knew his physics. Eventually one of these conversations derived into Dirac's monopoles and the enhanced symmetrical beauty of Maxwell's equation that would result if they were found. In this regard I said: "Unfortunately, Dirac left us, not knowing whether he was right." Lamb smiled and replied: "Well... he now knows."

On my return to Rochester, I received a kind letter from Willis Lamb and a few of his papers on quantum measurements (Lamb, 1986, 1987, 1989). I corresponded with him sporadically, on this topic, throughout the 1990s. My interest in this subject paid off years later when I used one of van Kampen's postulates to explain how cumulative single-photon quantum mechanical events gave raise to experimental observables in the macroscopic domain (Duarte, 2004).

Willis Lamb died in May 2009. In a small tribute I wrote: "A gifted experimentalist, and theoretician, in the best Newtonian tradition, he had a deep understanding of optical physics and the physics of optical measurements. His contributions to quantum measurements, and elucidative teachings on quantum mechanics, have not yet received the attention they deserve" (Duarte, 2009).

After *Lasers '91* Kathy left the organization of the conference and once again Vince began to deal with the logistics. By *Lasers '95*, held at Charleston, South Carolina, things were not going well and I got a small group of insiders to make an offer to Vince in an effort to gain control of the whole conference. Unfortunately, we could not reach an agreement and, at the same conference, I opted to leave. A few years later the conference folded.

6.3 Moscow

Following *Lasers '91*, I got an official invitation to visit Moscow. The visit was scheduled to coincide with *The International Conference on Advanced Laser Technologies* scheduled for September 8-11, 1992. I arranged my departure for early September. Prior to this trip, Ray Conrad, who had just been there, game me some valuable advice.

First, he told me not to drink the water, second, he described the long delay at arrival at the Moscow Airport, and thirdly, to watch out for cab drivers some of whom took westerners to the outskirts of the city where they were robbed and beaten. I also I got a phone call from a well-known physicist called Peter Franken, from Arizona, who knew Russia well and warned me that two weeks in Moscow was too long for a first visit. During this time, the Soviet Union had just collapsed, the presidency of Mikhail Gorbachev had come to an end early in 1992, and post-Soviet Russia was emerging under Boris Yeltsin. Undeterred I continued my travel arrangements.

I flew in United to Paris, on Tuesday September 1, 1992, and the next day I flew in Air France to Moscow. In the Air France flight I sat next to a young French lady. Reaching back to my high school French I began a conversation with her. It turns out that she was working on a documental film on Russia due to be released in a few months. Her name was Elizabeth, Lise for short. As we approached Moscow, and after some wine and a few laughs, Lise told me that a lady was waiting for her at the airport and offered me a ride to my hotel. At the moment, I did not think much of it and I had no idea of the importance of the offer Lise had just made.

At the airport, soon the reality of Ray's warning became apparent. An endless line of arriving passengers extended as far as the eye could see. At that moment Lise told me to stay close to her as we advanced parallel to the queue of tired and weary passengers towards the end of the line, towards the guards. There, a lady dressed in black waived to Lise who waived back. She was relatively tall, and slender, with dark hair. She appeared to be in her late 30s or early 40s. Her name was Julia.

Soon after greetings were interchanged, Julia took our luggage tickets and gave it to a man who returned with the bags, minus one. Then, it was arranged for a form to be completed by me so that I could recover the item later. Next, in complete disregard for the long line of passengers, we proceeded to exit the airport without any of the luggage being inspected by the guards. By any standard that was an impressive performance and obvious display of power. By now, there was no doubt that Lise's friend was no ordinary lady.

As we came out, two young physicists from the General Physics Institute (GPI) were waiting. One was Sergei Popov and the other was called Dima. At that stage, I proceeded to thank the ladies and told them that I had to go with Sergei and Dima. We interchanged phone numbers and went our separate ways. It was early afternoon and we proceeded to drive straight to GPI. Later in the afternoon, a phone call to the office of Victor Apollonov told us that the luggage had arrived. Dima volunteered to go with me to the airport and then to drop me at the hotel.

The drive to the airport on Dima's car turned into a three hour ordeal with the car overheating every few kilometers. This was however, an excellent tour of the outskirts of Moscow in which a never ending array of massive apartment complexes intermixed with sad human figures walking by the roadside under some unseen weight of hopelessness. At sunset, this bizarre scenery was given a fitting background by the setting sun which illuminated an orange horizon. Those long shadows of hopelessness were engraved in my memory. Eventually we made it to the Uzkoye Hotel. Dima went on to explain that this hotel was one of the best around since it used to be a former retreat for bosses of the Communist Party. The Party issue was brought to reality when they "kept" my passport at the front desk.

The hotel provided a small one bed room with a private bathroom. There was also a restaurant. Breakfast was: boiled eggs, bread, tea, some cheese, and Russian cookies. Next morning, Sergei Popov came to pick me up. The routine became one of visiting the various research projects at the GPI and talk to the researchers some of whom had to come from home were they were working on their private gardens. Although the reality at GPI was one of utter poverty and decay I treated my hosts with the up most respect. I knew of their contributions to laser science from the published literature and their names were familiar.

That evening, dinner was organized by Julia. Who was Julia? She said she was the daughter of a former KGB general. She also said to be part of an old Russian family from St. Petersburg. I had no reason to doubt her. She moved and commanded others like royalty. As time went by, she became my "protector." I did meet a couple of times with Lise, but, it soon became apparent that she was not part of Julia's schedule. Also, soon the rebel in me abandoned the official GPI plans and began to spend time at the gracious mercy of my new host who initiated a detailed and instructive series of visits to parks, exclusive shops, art museums, Communist Party museums, KGB exhibitions, Russian folk dance functions, the Bolshoi Theater, and well selected restaurants. One afternoon near her apartment, a few blocks from Red Square, she asked me to look at a building across the street "What do you see?" she asked. "A row of windows (6th floor) are bigger than the other windows of the building…" I replied. "Very good. That was Brezhnev's apartment" she concluded. Next, she showed me her school and the place where the feared Beria, a notorious former KGB boss, used to live.

The first Saturday we went to Moscow's street market which was quite large. There, merchants as well as ordinary people, were selling antiques, books, furniture, etc. I bought some fossils and a beautiful large marble apple among other things. As we were looking at some antiques, no further than 10 yards away, a couple of men wearing black leather jackets approached one of the merchants. The merchant was then shot point blank on the chest. A callous murder in broad day light and no one got involved. Worse, things kept going on as if nothing had happened. Apparently, we had witnessed the Russian mafia in action. Back in central Moscow, after dinner around 10 pm, Julia stopped a taxi and asked the driver to take me to the Uzkoye Hotel. They interchanged a few sentences and Julia told me that the driver was a technician at the GPI. Now I knew that the driver knew that I was a visiting Western scientist. Bloody fantastic I thought! I had already a mental picture of the way from central Moscow to the Uzkoye. Soon I realized the driver was not taking me via the usual avenues and remembering Ray's warning I began to prepare for the worse. Sitting behind the driver, in silence, I took the wide strap from my black thick denim fabric camera bag, and quickly made a plan to neutralize the threat. Looking through the rear window I could see behind, the lights of Moscow. I had no intention of being beaten or victimized. Thinking in clear and resolute terms, without hesitation or emotion, I calmly went over my simple plan... and waited. Then, the taxi took a sharp turn and immediately I saw the entrance of the Uzkoye. Pointing at his wrist watch, and smiling, the driver proudly gestured that he had taken a short cut. "Thanks God," I thought, and paid him double.

As I went into the hotel I was still not myself and I was completely startled at the resolute attitude of survival which had just taken over me. My adrenaline was still pumping and I could not go to bed. I headed to the bar for a drink. I needed a drink. There, I met an American "businessman" who proceeded to introduce himself. He was about my height, relatively slender, and gave the appearance as if in his fifties. He went on to explain his presence in Russia as looking for business opportunities in what could be a new frontier. However, he soon began to list a litany of unflattering characteristics of the society he discovered: utter corruption, crime, and deceit. Then, after a few more drinks, and in a sincere tone he confessed how he would wake up at night sweating with nightmares about what might happen to him. By now my recent experiences were relegated to the past and I said to him: "It seems to me that you must have more powerful reasons to be here than just business." There was no reply, as we both started to laugh. I never saw him again.

Next morning, Julia picked me up and drove me to a little town called Zagarsk. This was a church city. As I entered the main church my attention was caught by sun rays entering at the dome. As I began to focus in the high ceilings a spectacular display of light rays became apparent. Then all of the sudden, all the air around me began vibrating due to the powerful voices of the hidden monks. The singing was so powerful that I even felt the sound waves on my skin. Next, the women in the church joined in with harmonious high pitch voices. It was a display of acoustic beauty that induced an unusual and rare response on my senses.

We drove back to Moscow as I was still in awe. What I had just witnessed was a privilege, I thought. After more than seventy years of brutal Marxist tyranny, and persecution, the faith in Russia was still alive. Without any link to this culture I felt for the first time a slight nexus of belonging. This had been a most enlightening and spiritual experience and I was sincerely touched. In Moscow we stopped for lunch at a restaurant at the edge of Moscow's river. Julia told me that Raisa Gorbachyova was part of the "joint venture" that owned the restaurant. We spent the rest of the afternoon walking trough the main Moscow Park.

On Monday, Sergei Popov told me that the bosses at GPI, including Apollonov, were quite annoyed, almost furious, at my independence. He also said that Apollonov, and my American colleague, were going to "have a word" with me. Now, the conference was about to happen and we were due to move to Zelenograd. I refused. I told my American colleague that Uzkoye was "the best bloody hotel around" and that we could go, give our papers and still remain at Uzkoye. About two hours of push-and-pull went on. Finally, I accepted the pleadings from my colleague and agreed to take our luggage to Zelenograd. Sergei put my luggage on his car and parked it right at the entrance of GPI. The plan was to go to Zelenograd that evening. As I went into GPI, I asked Sergei: "Are you sure the luggage is safe?"

That morning I visited the copper laser laboratory as I was particularly interested in experiments on image amplification. About midday I came out of the lab to find Sergei in a desperate move waving his arms telling me that my luggage had been stolen right in front of the guards! Fortunately, I had with me all the important items, and that afternoon I went to buy some Russian underwear that I still keep, as souvenir, until today. I also went with Sergei to the police, as encouraged by GPI authorities. After observing humans in metal cages I said to Sergei: "Lets get the hell out!"

That night I had dinner, at Victor Apollonov's apartment, with Nikdi Cheburkin. Three things are worth mentioning. Going up to the apartment we were delayed by some young ladies curious about Western men. I was also surprised by the relative luxury of Apollonov's lifestyle that included a spacious apartment near central Moscow plus generous portions of *Le Cognag de Napoleon Courvoisier*. Cheburkin was obviously interested to learn about funding research procedures and he openly expressed annoyance at the lack of support he had encountered from US government laboratories. I don't know what, if anything, Cheburkin was told about me but we struck what seemed as an immediate and cordial friendship. However, after that dinner I never heard from Cheburkin again.

Later that night we had to deal with Zelenograd. That is the place where the Russians stopped the German troops in World War II. The conference place in Zelenograd was no Uzkoye. This was a dark cold place. It had long obscure corridors and a heavy, nasty, atmosphere. The rooms were bare concrete. That night I slept with my clothes on, including my boots. Next day, breakfast: a hard piece of dark meat… only God knew what from, yellow grease milk, some bread and tea. I drank the tea. At the table there were some tough looking Americans, from Germany, who seemed quite at ease at consuming that rubbish, they even seemed to enjoy it.

Immediately after "the breakfast" I told my American colleague: "I am getting out of here." Sergei drove me back to Moscow. There the issue became to find a place to sleep. The Spudnick Hotel was brought up as an alternative. The boys said: "It is a Party hotel" and the GPI has "inside connections" there. Thus, Sergei and I showed up at the Spudnick only to be told: "Sorry no vacancies." Sergei was not about to give up. They knew that a letter from the great Alexander Prokhorov (Nobel laureate) would get me in. Prokhorov was away but some how one of the secretaries produced a letter signed by Prokhorov requesting that I be given a room at the Spudnick Hotel. The letter was kept by the hotel manager. That same afternoon I moved in. That night, Julia took me to one of the best restaurants in Moscow in an attempt to compensate for all the recent troubles. At dinner, she ordered a couple of bottles of their finest vodka. I drank a whole bottle myself , and when I eventually went to sleep, I did so with the thought that next morning I would wake up with a most horrible headache. It didn't happen, apparently, she did order their best vodka.

Sergei Popov was quite upset and irritated at the theft of the luggage. We both suspected that it was not just a simple theft. He came to pick me up next morning at the Spudnick and proceeded to tell me of some new solid-state organic dye laser material. He then arranged a meeting with the professor in charge. This fellow wanted quite an elaborate research agreement and I had to explain to him that it was not possible. Sergei was observant throughout all this and once we were alone he explained to me that the professor was only a middle man. That was the beginning of my contacts with the scientists at the Medpolymer Institute in Moscow. More about this later.

I guess by now, Apollonov, and the others bosses at the GPI, gave up on me. I spend the remaining time under the protection of Julia. However, I went back to Zelenograd to give my paper on Friday morning. There, I was surprised to see the Americans from Germany still enjoying themselves at the conference. Obviously, they were very well trained. That evening, Julia asked me to meet a couple of charming Russian aristocrats. They were a banker named Boris and his wife Natasha. They also had a US based businesses in New York City. That evening, we all went to have dinner to a Club frequented by the Russian elite. There, I was introduced to the grand son of Nikita Chrushev and other Russian celebrities. Then, Natasha came up with the idea of going to one of Moscow's new night clubs. I was certainly not keen at this suggestion, but went along not to disappoint my hosts. We all went riding on Boris's brand new Mercedes.

It was Friday night. I was scheduled to fly back Saturday morning. However, there I was at a Moscow night club. Here, the police acted as private security guards. Every one was searched for weapons, and my camera was temporally confiscated. They openly displayed submachine guns and inside the club, men with baseball bats circulated from table to table. Never mind the naked dancing girls. All I could think at the time, was how to leave the stupid place. Eventually we got out.

About six in the morning, Sergei came to the Spudnick and we drove to the airport. There, Julia was waiting. There was an enormous queue of people trying to get out. They were Russian Jews leaving for Israel. I began to doubt my chances. But again my protector and facilitator went to work and after a polite good bye kiss I was pass Russian customs and security. There were loud and angry shouts from the crowd, but it did not matter. I was in the plane, again in Western hands. When I got to Frankfurt a member of the United crew asked me: "Where have you been?" And she kept on bringing me orange juice which I would quickly drink. During my Moscow visit I did not drink water, I was dehydrated, and had loss about 10 pounds. Otherwise, I was in good shape.

Back in the USA I got many long phone calls, throughout the 1990s, from Julia whom eventually left Russia and married a movie producer in California. However, she became a widow. Natasha called me several times from New York City and she even suggested a visit to Rochester. I gently dissuaded her from that idea. She had divorced Boris and was apparently looking for company. Sergei is now a professor at the Royal Institute of Technology, in Sweden, and we remain good friends.

By far, the most important outcome of my trip to Moscow was the contact made with the material scientists from the Medpolymer Institute. It was those material scientists the ones who synthesized and fabricated organic polymers for me later, under my own specifications. It should be said that we paid directly for these materials, via a British bank, in a straight forward commercial transaction.

These were the materials I used in the first near single-longitudinal-mode solid-state dye laser oscillators (Duarte, 1994). And this development led to further improvements in solid-state dye laser performance (Duarte, 1997, 1999). It is interesting to note that a discovery made in the US, in the late 1960s, was buried in the archives of knowledge only to be rediscovered, improved, and perfected by Soviet scientists... thus leading to further American innovation. That's the synergy of science.

6.4 Reflection

As time went by, a couple of things became clear to me. First, it was almost certain, albeit never discussed, that the US Army efforts on tunable laser development originated as a result of Soviet advances in this field. Secondly, it could be argued that the *Lasers 'XX* conferences, served as a vehicle to send messages, to the Soviet establishment, that the US was committed to high-power laser development. Thus, it is possible that this interchange, among laser physicist, in some measure, facilitated the end of the Cold War.

References

Baltakov, F. N., Barhikin, B. A., and Sukhanov, L. V. (1974). 400-J pulsed laser using a solution of rhodamine-6G in ethanol.

Duarte, F. J. (1988) (ed.). *Proceedings of The International Conference on Lasers '87*, STS Press, Mc Lean, VA. Duarte, F. J. (1991). *JETP Lett.* **19**, 174-175.

Duarte, F. J. (1991). Dirac: a scientific biography (Review). *Opt. Photon. News* **2** (3), 56.

Duarte, F. J. (1994). Solid-state multiple-prism grating dye-laser oscillators. *Appl. Opt.* **33**, 3857- 3860.

Duarte, F. J. (1997). Multiple-prism near-grazing-incidence grating solid-state dye laser oscillator, *Opt. Laser Technol.* **29**, 513-516.

Duarte, F. J. (1999). Multiple-prism grating solid-state dye laser oscillator: optimized architecture. *Appl. Opt.* **38**, 6347-6349.

Duarte, F. J. (2003). *Tunable Laser Optics*. Elsevier-Academic, New York.

Duarte, F. J. (2004). Comment on "Reflection, refraction, and multislit interference." *Eur. J. Phys.* **25**, L57-L58.

Duarte, F. J. (2009). W. E. Lamb. *http://www.opticsjournal.com/welamb.htm*

Duarte, F. J., and Conrad, R. W. (1986). Single-mode flashlamp-pumped dye laser oscillators. *Appl. Opt.* **25**, 663-665.

Duarte, F. J., and Conrad, R. W. (1987). Diffraction-limited single-longitudinal-mode multiple-prism flashlamp-pumped dye laser oscillator: linewidth analysis and injection of amplifier system. *Appl. Opt.* **26**, 2567-2571.

Duarte, F. J., Costela, A., Garcia-Moreno, I., Sastre, R., Ehrlich, J. J., and Taylor, T. S. (1997). Dispersive solid-state dye laser oscillators. *Opt. Quantum Electron.* **29**, 461-472.

Duarte, F. J., Davenport, W. E., Ehrlich, J. J., and Taylor, T. S. (1991). Ruggedized narrow-linewidth dispersive dye laser oscillator. *Opt. Commun.* **84**, 310-316.

Duarte, F. J., Ehrlich, J. J., Patterson, S. P., Russell, S. D., and Adams, J. E. (1988). Linewidth instabilities in narrow-linewidth flashlamp-pumped dye laser oscillators. *Appl. Opt.* **27**, 843-846.

Duarte, F. J., Ehrlich, J. J., Davenport, W. E., and Taylor T. S. (1990). Flashlamp-pumped narrow-linewidth dispersive dye laser oscillators: very low amplified spontaneous emission levels and reduction of linewidth instabilities. *Appl. Opt.* **29**, 3176-3179.

Duarte, F. J., Ehrlich, J. J., Davenport, W. E., and Taylor, T. S. (1993). A new tunable dye laser oscillator: preliminary report. In *Proceedings of the International Conference on Lasers '92.* (Wang, C. P., ed.). STS Press, McLean, Va., pp. 293-296.

Duarte, F. J., and Harris, D. G. (1992) (eds.). *Proceedings of The International Conference on Lasers '91*, STS Press, Mc Lean, VA.

Duarte, F. J., Taylor, T. S., Costela, A., Garcia-Moreno, I., and Sastre, R. (1998). Long-pulse narrow-linewidth dispersive solid-state dye laser oscillator. *Appl. Opt.* **37**, 3987-3989.

Kaiser, D. (2005). *Drawing Theories Apart: The Dispersion of Feynman Diagrams in Post War Physics*, University of Chicago, Chicago.

Lamb, W. E. (1986). Theory of quantum mechanical measurements. In *Proc. 2nd Int. Symp. Foundations of Quantum Mechanics*, Phys. Soc. Japan., Tokyo, pp. 185-192.

Lamb, W. E. (1987). Sequential measurements in quantum mechanics. In *Quantum Measurement and Chaos*, Plenum, New York, pp. 183-193.

Lamb, W. E. (1989). Classical measurements on a quantum mechanical system. *Nuclear Phys. B (Proc. Suppl.)* **6**, 197-201.

More, R., and Teller, E. (1988). On the chemistry of superconductivity. In *Proceedings of the International Conference on Lasers '87* (Duarte, F. J., ed.). STS Press, Mc Lean, VA. pp. 26-36.

Sze, R. C., and Duarte, F. J. (1989) (eds.). *Proceedings of The International Conference on Lasers '88*, STS Press, Mc Lean, VA.

Wang, C. P. (1986) (ed.). *Proceedings of the International Conference on Lasers '85*, STS Press, Mc Lean, VA.

Chapter 7

Eastman's

7.1 Introduction

Over the past century, American industry has produced some superb science and technology. Important, indispensable elements of our civilization are derived from these developments. The light bulb, telephones, photographic film, cinema, television, modern aircraft, the transistor, and the laser, are all products of industrial research in the United States. In the case of the transistor, further applied research led to integrated circuits, and this development eventually led to the computer, and computer networks made possible the internet. Additional semiconductor developments, in the US industry, also led to digital phones, digital detectors, and hence digital imaging. Of course, the fundamentals were pioneered at places of learning and universities. For instance, Maxwell's equations are essential to most of these devices and quantum mechanics provided the bases for the transistor and the laser. This is the essence of applied research, the use of fundamental knowledge to produce practical devices for the good of mankind.

The United States was very fortunate to count with visionaries such as Edison, Bell, Boeing, and Eastman. These men, under the freedom and protection granted by the US Constitution, created the conditions necessary to foster and encourage innovation and creativity in industry. This vision would ensure an unrivaled technological dominance for a long time.

This was more or less the case when I became familiar with industrial research and development in the US in the mid 1980s. Tremendous wealth, almost unlimited resources, support and encouragement from management were the norm.

In this Chapter, factual information is supported by references while the rest of the material reflects personal recollections and the opinion of the author. As such, this is only an approximation to a complete account on the totality of the events that led to the decline of this sterling industrial icon. Most likely, other authors will provide comprehensive accounts on the business side of things.

7.2 Arrival

A short distance from Macquarie University stood the newly built complex of the Division of Applied Physics of the Commonwealth Scientific and Industrial Research Organization (CSIRO). Impressive buildings in the heart of one of Sydney's most beautiful areas: Lindfield. Within it: physics, all sorts of physics. As a student I would visit their library and wonder about the prospects of working in such a place. A physicist's paradise, I thought.

When I was offered a job at the prestigious Photographic Research Laboratories of the Eastman Kodak Company, the first thing that attracted me was the salary. Their offer was more than double the amount I had been offered as an associate professor plus a signing bonus. This was a somewhat easy decision albeit there was a slight sadness at the thought of leaving academia.

Up on arrival, I was set up with a spacious laboratory with flat black walls and two large optical tables. The laboratory was promptly equipped to my specifications and taste. At the time, managers were all well-known senior scientists in their respective fields. The managerial structure tended to be horizontal so that they were relatively few and wise. I still remember with fondness the physicist who hired me: J. A. Merrigan. My immediate lab head was also a physicist, J. C. Kinard, a Southern gentleman who was extremely knowledgeable in the optics and technology of imaging. I also remember the impression that made on me the wonderful research library which subscribed to every journal that I could think of and the friendly and pleasant librarians, always smiling and eager to help. Jane Donaldson was particularly useful with searches and compiling reference lists for my first book *Dye Laser Principles* (Duarte and

Hillman, 1990). Besides the main research library, there were at least three additional libraries at different sites in Kodak Park.

Beyond the boundaries of research, Kodak Park was gigantic. It occupied more than 1000 acres in the North West part of Rochester, New York. It generated its own power, had its own medical facilities, its own fire department, and it was serviced by various railway lines. Parking lots were enormous. In business since 1880, by 1985, Eastman Kodak employed more than 60 thousand people in Rochester alone. Some three thousand of these were part of the local R&D community.

From my arrival in 1985 to 1992 my assignment was pure research. I was given some broad guidelines and some areas of interest, in the field of imaging, but it was up to me to choose the tools and the approach to solve the problems at hand. Besides, as long as I demonstrated progress in the assigned tasks there was complete freedom to pursue any interest in lasers and optics. If we add to this, the legendary generosity of Joe Merrigan, the division director, and the fact that Jim Kinard considered expenditures below ten thousand dollars as "noise"... then I concluded that I had found a paradise, of sorts, in the northern hemisphere.

Note: the Photographic Research Laboratories were founded in 1912 with Dr. C. E. K. Mees as director. For a historical account see James *et al.* (1989). Also a biography of George W. Eastman is given by Brayer (1996).

7.3 A physicist's observations

Albeit physicists from the Research Laboratories of Eastman Kodak were central to the creation of the *Optical Society of America* in 1916 (Hyde, 2006), and although the company had a very strong background and business in optics, chemistry was so pervasive that I always thought of Eastman Kodak as a chemical company. A diversified chemical company, where physicists were allowed to work on optics and other related fields. It should also be mentioned that when I came to Rochester, there was even a nucleus of professional mathematicians with some excellent Ph.D.s specializing in numerical analysis.

Prior of my arrival to Eastman's, I knew of their laser work, I knew of the invention of the continuous wave (CW) dye laser, I knew of the work of O. G. Peterson and colleagues. Later, I had the pleasure of working with one of the contributors to that effort, Harvard Ph.D. physicist, J. P. Webb. Pierce was a senior physicist, son of Julian Webb a former Director of the Research Laboratories who is mentioned in one of Feynman's books (Feynman, 1985). A clever and widely applicable invention, such as the CW dye laser, was not transformed into a product. This was not an isolated case. In many and long conversations, with Pierce and other old timers, I was introduced to the idiosyncrasy of the place. From these conversations plus my own observations soon I developed a simple set of guidelines:

(1) Don't seek direct participation in "hot projects" since there was a good chance that these were politically charged engineering tasks.

(2) Publish in the open literature. This often meant working on long term research with no immediate commercial value.

(3) Find an area of expertise where you could contribute directly, routinely, and substantially, to the company's bottom line. This would make you useful. The area I found was optical metrology.

Opinion on commercial value: the more original the research is, the farther away its commercial realization. In my case, the line of tunable laser research could have been very valuable and was indeed valuable to outside colleagues reading my papers. However, the company was so rich that apparently managers had already decided, prior to my 1985 arrival, that lasers offered limited commercial or immediate monetary value. And arguably, relative to film, this perspective was true, at the time. Thus, lasers were a perfect area of endeavor where one would be allowed to publish. As a physicist, my duty to the company was fulfilled by solving production problems, improving instrumentation, advising colleagues when required, publishing in refereed journals, and submitting patent applications when appropriate. I decided very early not to use my time trying to sell my research. That is, I had made the decision to be a physicist, not a salesman. This was a pragmatic decision.

On open literature refereed publications: perhaps in order to protect company information, or to encourage patenting, it was not easy to publish. Every paper submitted to a journal for refereeing was required to go through an approval process which necessitated a litany of managerial and legal signatures. This varied from organization to organization, but, at one time I counted seven approval signatures attached to one of my papers. A snag anywhere in the chain of approval could condemn a work of science to the archives of oblivion. It took persistence and stamina to publish. Unfortunately, many colleagues became discouraged by the system and eventually would give up on publications.

From my days as a student, postgraduate, postdoctoral, and as a professor, I was used to work Saturdays and Sundays. I continued that ethics at Eastman's, but soon realized that I was almost alone in this practice as most stayed home during weekends. This conveyed to me that most people assumed the privilege of scientific practice as a job rather than as a vocation. The enthusiasm was absent.

Soon after my arrival, I was identified as a physicist removed from the mainstream. Some did refer to such characters as "individual contributors." This is someone who worked in a given field and contributed to solving problems encountered by R&D efforts often associated with a line of products. In my case, enough time was left to focus on my beloved physics. As hinted previously, my business contribution was in the area of optical metrology for photographic and cinematographic films, plus other imaging and optical products. The company in the late 1980s and early 1990s was healthy and strong enough, to allow for a select number of such individuals, albeit the value of nimble strategic research efforts was not fully understood or appreciated. Problems brought to me, by engineers and managers, were always promptly and effectively solved.

7.4 Laser physics at Eastman's

Students, and friends at large, would often enquire on my work as a physicist. For a physicist, work flows from ideas. In this regard, John Ward always said that a good physicist only had "a few truly original ideas" in his lifetime. In principle, that appears to be correct. However, a few original ideas might generate a number of variations and applications. Here, I describe some of these ideas, relative to my tenure at Eastman's, with a brief explanatory note on some immediate applications. This list covers the 1985-2006 years:

1. The introduction of the generalized multiple-prism dispersion theory for pulse compression (Duarte, 1987, 1990). Applications: Laser pulse compression, design of ultrashort (femtosecond) pulse lasers, and microscopy.

2. The introduction of broadband multiple-prism polarization rotators (Duarte, 1989a). Applications: laser beam propagation and laser system design.

3. Demonstration of lasing in coumarin tetramethyl dyes (Duarte, 1989b). These laser dyes became a rapid commercial success. Applications: lasers for spectroscopy and micro lithography.

4. The first matrix description of multiple-prism beam expansion and multiple-prism dispersion (Duarte, 1989b). Application: optics.

5. The introduction of *N*-slit laser interferometers incorporating digital detection (Duarte and Paine, 1989; Duarte, 1993a, 1993b, 1995a). This invention included the first demonstration of illumination with a very thin expanded beam of light (or light sheet) of a sample for microscopy purposes. Detection was performed using a digital sensor such as a diode array or a CCD detector. Applications: interferometry, optical metrology, microscopy, imaging, characterization of optical films and surfaces. (Note: this work benefited from the excellent computer programming of David Paine.)

6. The introduction of Dirac's notation to the generalized description of *N*-slit interference (Duarte and Paine, 1989; Duarte, 1993a, 1993b, 1995a). Applications: imaging, optics, and optical design.

7. The introduction of tunable solid-state narrow-linewidth dispersive laser oscillators (Duarte, 1994, 1995b, 1999). Application: laser spectroscopy and photochemistry.

8. The unified description of interference, diffraction, refraction, dispersion, and reflection via the generalized interference equation derived using Dirac's notation (Duarte, 1997, 2003). Applications: optics and laser propagation.

9. The design and construction of large-scale high-throughput ultra-fast industrial laser printers (Duarte, 2001; Duarte *et al.*, 2005b). Applications: laser printing and imaging paper sensitometry.

10. The demonstration of secure free-space interferometric communications (Duarte 2002, 2005). Applications: optics communications and laser propagation.

11. The introduction of tunable solid-state organic lasers incorporating functional distributions of nanoparticles in the gain medium (Duarte and James, 2003). Applications: laser design, laser spectroscopy, and photochemistry.

12. The first demonstration of coherent emission from electrically excited organic semiconductors (Duarte *et al.*, 2005a). Applications: miniature coherent illumination, OLED coherent illumination.

13. Derivation of positive and negative refraction laws (also known as Snell's laws) using quantum principles. Multiple-prism dispersion equations for negative refraction (Duarte, 2006). Applications: optics, unknown future uses in laser optics.

In summary, my tenure at Eastman's is documented in numerous conference papers, more than sixty refereed-archival publications, six books, three conference proceedings (as editor), and only nine US patents. This type of record was somewhat atypical since many other colleagues would mainly publish large numbers of patents.

During this time, I also received several professional recognitions from prestigious institutions such as the *Australian Institute of Physics* and the *Optical Society of America*. On a more personal note, in 1987, Diana was born.

7.5 Early departure

Joe Merrigan was an extraordinary character. In the summer of 1986, I figured out that I needed a excimer laser to complete the infrastructure of my laser laboratory. After talking to friends and some searching, I settled for a powerful prototype that a company agreed to provide for about 40 k. All I needed now was the money. Thus, I decided to call Joe Merrigan at around 5:30 in the afternoon. He was still in his office and said: "Well, come and explain to me how are you going to use this laser." Soon after hanging up the phone I went running to his office which was about half a mile from my lab. When I got there, I was sweating. He smiled, and said: "Well, since you came running, here is my signature." Besides some fun experiments, that laser made a lot of money for the company via the demonstration and testing of new laser dyes.

In the spring of 1988, something happened that would have a profound effect on the rest of my career at Eastman's. However, I did not suspected it at the time.

One afternoon, around 6 pm, I was going home as it was raining quite hard. My 1970 Ford Ltd was parked far away and I was looking through a corridor window contemplating my options. It was then that I saw Joe Merrigan, via the reflection of the window. Smiling, and in a soft voice, he said: "Frank, let me give you a ride."

From my friend Paul Wojciechowski, I knew that he was ill but said not a word about it. As we drove, he asked about my research and I explained to him some of my latest experiments and as usual he smiled and exclaimed: "*Hallelujah!*" He stopped next to my Ford Ltd and smiled for the last time. Joe Merrigan left this world a few weeks later. With his departure I lost hope of having a corporate mentor. He was a bright, highly ethical, decent, and generous man, a good physicists, and a strong advocate for scientific innovation at the higher echelons of the company.

7.6 Initial changes

Competition from abroad combined with an emerging MBA culture began to exert their influence. By the late 1980s, well-qualified and experienced managers were encouraged to retire with highly attractive separation packages. The company had began to ignore its own human capital: a power vacuum was created.

The horizontal managerial structure was replaced by vertical silos, and a new short-term business dominated culture began to emerge. The new, far more numerous, managerial positions were filled mainly with business graduates, engineering-business graduates, and non-practicing scientists. In my opinion, as implied by Schrödinger's prophecy (Schrödinger, 1992), the scientifically and technologically challenged began to gain prominence.

Some of the first changes noticeable were a demand for attention away from work, away from experiments, and away from tangible projects, towards meetings and the adoption of new cultural concepts. Also, in an apparent effort to save money a few functions began to be performed by outsiders in what appears to be early versions of outsourcing.

The concept that a manager could manage anything, was now in vogue. This dogma had some serious consequences. Some unqualified, and scientifically challenged individuals, were allowed to lead complex technical businesses and organizations. When they first got there, things were running well, even like clockwork. However, to reduce cost they began to "let go" qualified scientists an engineers. The tool known as "reorganization" was used repeatedly in this context. Not even experienced technicians were spared.

In my opinion, this was somewhat analogous to a bloodless "cultural revolution." One of the various manifestations of what I called corporate Marxism, a kind of arbitrary benevolent-egalitarian transformation, where those with limited scientific qualifications were put at a par, and even surpassed, the learned and the knowledgeable. Most of my colleagues would assume a behavior of compliance, indifference, or neutrality. I did not.

Since 1992, until my exit in 2006, I witnessed a series of downsizings. In total, I went through about a dozen downsizings. Initially, these were events full of collective drama and suspense. But soon they became predictable rituals. At first, I was too young and arrogant to be concerned. Furthermore, as time went by I began to even welcome these ritualistic periods as they represented an excellent opportunity to work exclusively on my beloved physics.

Given my outspoken criticisms, and independent, almost rebellious behavior, it is not immediately clear how the system tolerated me as long as it did.

7.7 War and business

Of the many outside trends that became apparent in the American corporate culture in the late 1980s, was an interest in the Japanese culture. This was justified in this case since Fuji was perceived as the main competitor. Hence, around 1990, I was given the opportunity to practice the martial art of Iaido. This is the study of Samurai swordsmanship. Briefly, this was taught by a former Kodak scientist, Cynthia Jones (2nd Dan), who learned from a Vietnam veteran, John Drouin (3rd Dan), who had trained under Japanese sword master Takeshi Mitsuzuka. In 1993, I was promoted to black belt. Besides practicing with wooden swords called "bokken," and real swords, elements of the Japanese culture were studied. Among these was a little book written by a Samurai warrior, in the 16th century, called Miyamoto Musashi (1982). The book is called *A Book of Five Rings* and it is said to be studied, as a work of strategy, by Japanese industrialists and businessmen. A lot in that book is just plain commonsense. However, many managerial types have given lip service to this class of writings, and the ancient Chinese book *The Art of War*, by Sun Tzu (1996), to promote a kind of management where business practices are analyzed as some kind of war strategy. That is, competition is considered within the boundaries of a theater of war. To many, this might sound reasonable. However, this approach misses the all important fact that true successful businesses don't have competition. In other words, if you have a truly unique product you don't have a war to fight. The pioneers of American industry understood this well. Steve Jobs understood that well. That is why they created outstanding R&D efforts that would always maintain that truly unique competitive edge.

A *kata* is a series of predetermined steps, sword movements, and sword cuts. One day, after about my 5th year of training, a student asked me to demonstrate a particular kata given that he had some doubts about some of the sword cut. After a brief discussion of the movements I proceeded to assume the initial posture and to execute the kata. The sword left silently its scabbard (saya) and soon the Dojo became dominated by the sequential sounds typical of a succession of sword cuts, following the motion of an arc, in space. In the final downward cut, where the rotational movement of the sword is suddenly brought to an end, the sword broke... *it snapped,* and the blade continued as a projectile ending its motion firmly stuck in the hardwood floor barely one or two inches away from the feet of the student. There was absolute silence in the Dojo.

The value of my sword was only a few hundred dollars since it was made using a modern process and not under the exquisite craftsmanship associated with ancient samurai swords. Just beyond the boundary of the handle, at the habaki, there is a notch. Following thousands of cuts and violent motions, due to metal fatigue, the steel sword had cleanly snapped right at the notch. Observing the cross section of the sword the regions of high-carbon and low-carbon steel are clearly visible but in the absence of the high definition and geometrical boundaries so typical of highly valued swords. Metal impurities were also a possible contributing factor. No one there had ever seen such a break before. Sensei Jones suggested to call the vendor, a Japanese businessman, to complain and perhaps get a free replacement. His answer to my call went something like this:
"You do very dangerous sport... in Japan people killed *aaall* the time." Given that persuasive argument I had no alternative but to purchase a new sword.

7.8 The measurements business

Around 1991, I was briefly assigned to a new group. At the first meeting I had with my new leader, he showed me a neat flowchart depicting my new functions. I made a few comments about his flowchart and he became much annoyed. Soon, he left the meeting very angry, almost enraged. However, one day he invited me to give a talk. Surprised and delighted, I accepted. The talk was on a digital laser *N*-slit interferometer, I had just invented (Duarte 1993a, 1993b). However, unknown to me he saturated the audience with his friends. Also present, was Jim Terwilliger the successor to Joe Merrigan. As soon as I began to speak, questions began to be asked. I politely welcomed the interchange and proceeded to answer the questions, but soon it became clear that there was a hidden agenda, and for 45 minutes I battled part of the audience. Jim Terwilliger, in absolute silence. He had a Ph.D. in chemical engineering and apparently was unable to follow most of the discussion which was on optics. At the end of the talk, there was no applause and my host proceeded to say: "You know Frank, I didn't understand a word you said" to which I replied "Sorry you did not understand." Then he followed: "Sitting here I have several Ph.D.s and they did not understand eiiither." Smiling I replied: "You should send your boys back to school." There was an immediate and loud uproar in the room. Soon after the meeting the new leader declared me "unmanageable" and I was freed to leave and reintegrate my previous lab. Subsequently, he became ill and proceeded to retire early.

In 1992, Jim Terwilliger asked me to take charge of a laboratory that I named the *Imaging and Spectral Measurements* (ISM) Laboratory. This was the optical metrology laboratory, of strategic importance, that performed all the optical measurements for the film community at Kodak. Services included: spectrometry, microdensitometry, and modulation transfer function (MTF) measurements. I added laser and digital interferometry. The laboratory had some 25 people in it, mostly well trained technicians, plus a few engineers and scientists. No photographic film, or motion picture film, or imaging surface went on to production unless it was evaluated at ISM first.

At its peak, around 1995, ISM performed nearly *one million* measurements per year and did it so at an extremely competitive cost which was a fraction of the usual cost associated with external laboratories. Parenthetically, when I suggested to offer the ISM services to outside customers, my suggestion was met with a paternalistic no.

ISM was a very well thought-out laboratory. It was the result of a piecemeal evolution led by very clever physicists, and top optical scientists, over decades. Many of the instruments were designed and built in house. In that category, we had a family of high-speed microdensitometers (also known as "granularity machines") and an automated modulation transfer function (MTF) instrument. Next, there was an array of state-of-the-art spectrometers.

Soon after assuming my responsibilities at ISM, I realized that the managerial part could be taken care of, on average, with less than 25% of my time. The rest was used to work in my lab either doing research or instrument development. This was made possible by the skill and professionalism of many of the senior technicians including Glenn Gursslin, Mike Nentarz, Janet Whiteford, Jim Young, and Mike Zawacki. Computer engineering was under the supervision of Jim Hill, an experienced and capable man. A routine business function that I performed every day, on my arrival, was to sit down with one of the technicians, Karen Eisenhauer, and go over the control charts of some of the most sensitive instruments. If something was questionable, the instrument was immediately taken off line for observation or to perform the necessary adjustments and/or maintenance.

I assumed a very open style of management and I was very clear on my expectations to each individual in the laboratory. I also gave a course on optical measurements including assessment via a written exam. As a result, on performance appraisal time, I never spent more than five minutes with each individual and there were no disputes. Grateful film scientists would even ask Jim Hill and I to pose next to Oscar statues won by movie films measured at ISM. The business ran like clockwork and the research was fun. What more could I ask? However, winds of change were just around the corner.

By 1993, a major shift was set in motion. Due to a large reorganization our division (with about 600 people), headed by Jim Terwilliger, was included within a new and larger coating division.

Most of the new managers came from the production lines. At best they had a limited understanding of research. Soon we saw the new order being imposed by forcing lab heads to wear raccoon hats and calling themselves "the pioneers." Certainly, Terwilliger was quickly reassigned from division head to lab head and eventually displaced altogether.

One winter day, in 1994, we got a note on a Friday to urgently, and compulsorily, report next Monday to the main office building. There, we met an outside trainer, who could be described as a motivational speaker. By Monday lunch time, he had already revealed unnecessary details of his personal life. Apparently, the aim was to get those in the leadership ranks to reveal personal stories so that they would "bond" with the team. I steadfastly refused to participate and every time they gave me a piece of paper to write something I would write nothing, or I would write "confidential."

Nevertheless, the men participated. This went on all week long and by Thursday afternoon the full slate of "pioneers" went on stage and the motivational speaker announced: tomorrow we'll discuss the undiscussables and went on to offer the podium to the chief pioneer. He soon began to give a speech. By now, the tension had been slowly building and the men in the audience looked tired and alienated. As he was elaborating on a particular point I stood up and said something like: "That's not true..." Immediately, the men engaged in a chain reaction. First one, then a few, then many in the audience stood up expressing their grievances. As I left the room to get a cup of coffee, I could hear the collective shouts from the distance. On Friday, there was no discussion of the undiscussables. They had already been discussed, I guess. By then, I surmised that my future in the new organization was in doubt.

Regardless of my open opposition to the new culture, and to their credit, in 1995 I was invited to join a select group with the implication that it would be very good for my career. Incredibly, they had given me another chance. However, by then I had already arrived to the opinion that this new culture did not align with my ethos and that management, in this environment, was not my cup of tea.

By late 1996, I was engaged in almost daily skirmishes to preserve the integrity of ISM. Given the sterling business performance of ISM plus my standing as a physicist in the company, I could not just be moved out. Not easily, anyhow. Then my new boss, whom often complained about my lack of compliance, decided that I needed a career change and proceeded to reorganize me, and Jim Hill, out in what became a mini reorganization. In an astute political maneuver, he facilitated this reorganization saying that this would be good for my career, despite the fact that I was not consulted. Following my appeal to higher authorities, he was charged with the task of finding me a new job. This process took more than six months.

In the meantime, J. W. Meyer the then director of the Research Laboratories sent me on a glorious tour of Europe on a digital assignment. Late in February 1997 I departed to visit England, Monaco, and Spain. In England I visited the Clarendon Laboratory, at the University of Oxford, gave a couple of talks, and spent time with several laser researchers including Colin Webb (FRS), whom introduced Jim Piper to lasers. Colin arranged for me to stay at Jesus College, invited me to a formal dinner, with the college elders, and took time to drive me (in his new Jaguar) around the English country side. Whilst at London, I visited the research facilities, at Harrow, and femtosecond researcher Paul French, at Imperial College. At the Monaco airport Etienne Philiponet, a researcher from Kodak-Pathé, was waiting for me. There was a conference on digital technology in Monaco but we had lots of time to drive around. Etienne took me to castles and Roman ruins, either in France or Italy, and would always stop at the best possible restaurant. Needless to say, we struck a rapid affinity and spent time talking about physics and sports. I then asked him why was he allowed to spend all that time with me. Etienne answered that this was at the explicit request of his boss Nicolas Touchard. This was a bit of a pleasant surprise since I knew that Nicolas visited regularly some of my critics back in Rochester.

In Spain we visited the research institute "Instituto Rocasolano" where I gave another talk and spent time in long conversations with my friends Angel Costela and Roberto Sastre who were also doing research on solid-state dye lasers. Angel organized visits to several historical sites and took me to the area in Madrid where the great Cervantes (author of the immortal *Don Quijote de la Mancha*) lived. I had so much fun, in this visit to the old continent, that I almost forgot the purpose of my visit.

The joy of the trip was tempered when I received a phone call telling me that a new job as a laser physicist had been found. I was to work on industrial size laser printers. My new boss was John Pardo. John was himself a competent physicist who had been trained in high-power CO_2 lasers.

Following my departure from ISM, layoffs began and not all ended well for the technicians. One of them, a lady near retirement, after loosing her job, also lost her house where she lived with her elderly mother. In retrospect, as an unseen blessing, the reorganization had shielded me from participating in this type of managerial work.

Worse is what happened to Jim Young, the microdensitometer man. He was very bright but he was not very sociable. One day he came to see me, had tears in his eyes, and said: "Frank, they are making life very difficult for me." Jim was single and this job is all he had. I also knew that he had invested quite successfully in the stock market. I suggested to him to get his money and join his brother in California. He agreed. A couple of weeks later I came back to my office and found a letter that said: "Frank, thank you for every thing. In all my years here you were the only one who gave me a fair chance. Adios amigo, Jim." One week later they found him dead from a heart attack. He was an African American, in his early fifties, of slim built.

During my tenure at ISM some remarkable improvements in timeliness were achieved. This, plus external recognitions to my physics, provided the opportunity for some communications with George M. Fisher, the CEO at the time. For example, when we reported on a X20 cycle time reduction, in our high volume microdensitomer measurements, the CEO responded: "Fantastic progress!" (Fisher, 1995a). In a previous personal note he had stated: "Our company is stronger today due to your inventiveness..." (Fisher, 1995b).

In 1997, from my new position, I decided to confront those who had separated me from ISM. Thus, I wrote to George Fisher, and other higher managers, suggesting consolidation of measurements and coating facilities. Eventually, my previous division was dismantled and its various functions distributed among sister divisions . The ISM instrumentation was transferred into a section of my newly adopted division. Most of the leadership of the previous organization was retired.

A foot note: during this time, one of the nicest compliments I was paid was meeting a recently hired young Ph.D., from a leading Ivy League university, who told me that she had come to Rochester in the hope of working with me on tunable lasers. However, when she arrived it was realized that although she was hired to do optics she was in a different organization. Given some organizational barriers we could not work together. She quickly became disillusioned, and left the company. Remarkably, I also got a phone call from one of the physics students I knew at the University of Alabama. He now had a Ph.D. and was about to interview with the company. Choosing my words very carefully I encouraged him to look elsewhere.

7.9 But aren't you from Eastman's?

When I was still part of ISM (around 1995), a top manager sent a group of scientists to Latin America. We visited Mexico, Colombia, Argentina, and Chile. The main purpose of the tour was to teach digital imaging to the local employees and also to interact with the local customer base. In Mexico, we were very well received, by the customer community, and enjoyed some wonderful hospitality, particularly in Guadalajara. In Argentina, we only taught the course and visited some of the best steak restaurants in Buenos Aires.

In Chile things were a bit different. As soon as I arrived at my room, in a five star hotel in Santiago, the phone rang. A friendly young lady, in perfect English offered me a visit. I proceeded to politely decline the offer to which she countered by asking: "But aren't you from Eastman's?"

When I went down to the lobby she was still waiting and insisted. She was fairly attractive, dressed elegantly in black, and said her name was Gloria, but did not explain why she asked that peculiar question. I reiterated, as nicely as possible, this time in Spanish, that I was not a salesman, I was not a businessman, nor a manager. I was just a scientist, and would not require her company. Her surprise was even greater when she learned that I was originally from Chile, at which time she proceeded to withdraw almost apologetically.

In Chile, as in Mexico, we interacted with both the costumer community and the company's representatives. In both places the customers expressed specific grievances. On the conclusion of these travels we were verbally debriefed, back in Rochester, but no written report was requested. By now, it was clear to me that besides an inability to develop and manufacture successful products from new inventions, there were also some problems in the marketing of our existing sterling film products. I was glad that I was a scientist and did not have to deal with the business side of things.

Back at the labs: lack of consistency, reduced transparency, and poor decision making, were now hard to ignore. The promotion of mediocre and scientifically deficient individuals gave a perception of decline, and was particularly worrisome. Additionally, some of the scientifically challenged appeared not to observe the same ethical standards of people like Merrigan, for instance (more on ethics, later).

Going back to the laser dyes mentioned previously: the coumarin tetramethyl dyes, which I demonstrated as efficient laser media in the blue-green (Duarte 1989b), sold very well since they were also highly soluble even in water. According to Dick Hapeman, who led the laser dye business, a well-known industrial giant bought large quantities of these dyes to be used in micro lithography. Dick introduced a nice laser dye catalog late in 1987, where I wrote an introduction to dye lasers (Duarte, 1987b). This publication was very well received by the laser community. A second enlarged edition soon followed (Duarte, 1989c).

Business was very good and Dick was happy. However, one day, Dick was replaced by a young business lady. I kept working with the business lady, but it was clear, by then, that management did not believe in the long-term viability of organic laser dyes. The end to the well-known, and even iconic, laser dye product line came when the whole of Eastman's Fine Chemicals was sold. Thus, in a unilateral move, the laser dye market was left in the hands of our competitors, who continue to be in business.

7.10 Strategic moves

The genius of George W. Eastman created a great company. It was a true American icon. The company grew so strong and diversified that even with the disorder just described the company had a good chance of weathering the decline of film. However, new strategic moves would place its future very much in doubt.

In 1985, at is pinnacle, Eastman Kodak was a diversified chemical company. Diversification was such that during WWII, its scientists participated in the isotope separation effort for the Manhattan Project (James *et al.*, 1989). Albeit a photographic company initially, imaging became only one of Eastman's production emphases. In addition, the company produced a wide variety of materials, chemicals, optical products (Kodak, 1990a, 1990b, 1990c) and even bought the pharmaceutical company known as Sterling Drug. There was also an entire division dedicated to work on defense government contracts.

Following the managerial vacuum created in 1992, apparently the remaining upper managers lost confidence, began to redefine the business, and went outside for direction. First, they spun off Eastman's chemicals, which became the Eastman Chemical Company, and Kay Whitmore was replaced by George M. Fisher. He came from Motorola and had a mathematics background.

I never met Fisher but I communicated with him several times and in the people skills department he was a high-class act. However, in my opinion, three strategic mistakes were made during the Fisher era. The first mistake was to continue to redefine the focus of Eastman Kodak as purely imaging. This new definition neglected that the company was much more than imaging, this was a fundamental shift. Further significant steps in this imaging direction were the sales of Sterling and Bayer Aspirin, at a loss, in 1994.

Transforming Eastman from a widely diversified materials and chemical company to a narrowly focused imaging company was, in my opinion, the first and most significant error of the Fisher era. For a while, things did look quite good and the stock price got to be nearly 93 dollars, early in 1997. Due to his reduction doctrine, cuts and downsizing also became part of the new doctrine.

In addition to selling assets, Fisher also began a policy of investment in China. This was the second mistake of the Fisher era, although at the time it did seem like a good move. In 1999, I communicated with him criticizing the wisdom of this decision. With sincere idealism, or possible naïveté, he replied: "Frank, it would be awfully shortsighted for us to rethink our position relating to the opportunity in the Chinese market… the opportunity… is so immense we dare not turn away" (Fisher, 1999). I remained unconvinced, and eventually the China investment became a major loss. In retrospect, this move was one of the first significant steps in the shift of American manufacturing to Asia.

The third error of the Fisher era was the selection of his replacement: they opted to leave in his place a company insider, whom in my opinion, did not have the technological vision necessary to steer the institution towards a successful future. This was Fisher's first derivative (FFD). Following Fisher's departure, I lost all interest in larger company issues and focused instead totally on my work. By now, downsizings were routine and it was clear to me that the old grand Eastman Kodak was no more. By now, managerial terminology such as "six-sigma" and "black belt," was abundant.

Film versus digital: many colleagues, and commentators in the press, believe that the delay to embrace digital was what caused the company's decline. To blame digital technology for the failure is a misguided excuse. Many companies made a successful transition from the analog era to the digital age initiated by the invention of the CCD detector. These companies include Canon, Nikon, Leica, and yes, Fuji Film. Although failure to dominate digital technology added to the decline, alongside managerial decisions, I don't think it is as important as the first mistake of the Fisher era. Furthermore, it is relevant to highlight that a successful transition from film to digital would also necessitate a transition from a chemistry based culture to a hardware culture. As a matter of fact, in 1997, I communicated with Fisher cautioning him that electronic hardware was not a strength of Eastman's and that, as a chemical company, it would be preferable to concentrate on the materials side of digital imaging.

Was I qualified to issue such observation? I think so. From my Macquarie education I knew about the physics of transistors and I understood CMOS logical gates and I had designed and built small dedicated computers to handle the data from my laser experiments. Furthermore, at Eastman's I pioneered the use digital detection in *N*-slit interferometers designed for imaging diagnostics (Duarte 1993a, 1993b). A joke I used to tell on this topic was that, without changes, the company would only see the zeroes out of digital imaging.

Note: molecular imaging, or film technology, or photography, should be around for a long time, for specialized applications in the arts and sciences. This is due to its superior resolution and its unmatched archival qualities. The archival superiority of traditional black and white silver-halide films is an issue almost completely ignored by the mass media and the public. Also, molecular imaging protects the integrity and fidelity of the events being recorded, a feature that has advantages in scientific and legal applications. Parenthetically, it appears that Fuji Film will end up as the sole beneficiary of this American technology.

7.11 Transformation

In 2005, Fisher's replacement (FFD) was replaced by a former inkjet manager from Hewlett Packard, who arrived in 2003, and became Fisher's second derivative (FSD). The years 2004 and 2005 were pivotal. Up to this point there was sufficient optics, chemistry, and materials infrastructure left in the company to provide a successful future. However, late in 2004 the new management sold the historical optics business to ITT. A good business of extraordinary strategic importance was no more. The transformation continued, and early in 2007 they announced the sale of the iconic health imaging business, which was part of the company since 1895, to Onex. Late in 2009, they sold the organic light emitting diode (OLED) business to LG Electronics. By October 2011, the digital patent portfolio of the company became one of the last valuable assets left (Spector and Mattioli, 2011). By November 2011, they sold the digital sensor business (Reuters, 2011a). By December 2011, they sold the historic gelatin business (Mattioli and Fitzgerald, 2011). And by February 2012, in what might be interpreted as a display of defeatism, they got out of the digital camera business (Mattioli, 2012). Indeed, it might not be surprising if in the near future parts of the former Kodak Park are sold to extend the cemetery that lies just north of the park.

In 1985 Eastman Kodak employed more than 60 thousand people in Rochester. In the area of research and development alone the company was said to employ some 8 thousand people, world wide, by 1989 (James *et al.*, 1989). That was the pinnacle.

Decline: according to a public disclosure, 43 thousand jobs were eliminated in the 2003-2011 interval (Kodak, 2012). The total number of employees, in Rochester, had declined to about 9 thousand by early of 2008 (Daneman, 2008). Downsizings continued into 2009 (Daneman, 2009). More reductions brought numbers down to about 5 thousand plus by the end of 2011 (Daneman, 2012a). Additional cuts, that would take the numbers below 5 thousand were disclosed in 2012 (Daneman, 2012b). Thus, since 1985 the number of employees, in Rochester, has been reduced by more than 90%. A most definite trend, it seems. When an army goes into battle and looses 10% of its soldiers it is said to be *decimated*. Kodak's losses are far beyond decimation. They represent near obliteration.

In a 2005 FSD speech, the hypothesis of a "flat world" was used to support the vision behind the Kodak "digital transformation." The notion of a flat world is completely erroneous even in a metaphorical sense. The reality of a nearly spherical world was even accepted by Queen Isabella (Reina Isabel, in Spanish), in 1492, when she financed the exploration of the new world. Her Majesty might have even been inspired by the scriptures (Job 26:7; Isaiah 40:22) to discard the notion of a "flat world." Alternatively, like many others, she might have been familiar with Pythagoras's astronomical deductions (Vanderbei, 2008).

By the summer of 2007, the local press reported that nearly 80 building in Kodak Park had already been demolished (Daneman, 2007). This gave an ironic twist to the notion of a "flat world."

Early in 1997 the Kodak share was traded at nearly 93 dollars. This value steadily declined to about 28 dollars by mid 2005 and to the 15 dollar range by mid 2008. By the end of July 2010 the Kodak share traded just below 4 dollars, and by the end of December 2011 it was trading at 0.65 dollar, or 65 cents (see, for example, Google (2011a)). Yes, by the end 2011, the Kodak stock became a penny stock. This means that in 2011 alone Kodak shares lost about 88% of their value.

Kodak stock began 2012 submerged in the penny realm and facing bankruptcy (Spector and Mattioli, 2012). On January 19, 2012, Kodak filed for bankruptcy (Reuters, 2012) and trading on its shares was halted while they decayed to a value of 36 cents (Google, 2012a). By contrast the Eastman Chemical Company, spun off in 1993, traded by the end of that day at over 46 dollars per share (Google, 2012b). At the same time, Fuji Film traded at nearly 25 dollars per share (Google, 2012c). Overall: in the February, 1997 - April, 2012 era, the Kodak stock decayed by more than 99.5%.

A local work force reduced by more than 90% (1985 - 2012), the value of the stock evaporated by more than 99% (February, 2005 - April, 2012). Flattened buildings and bankruptcy: this is the sorry spectacle of a transformed, and prostrated, American icon. Mysteriously, FSD continued to hang on to the corporate wreckage even into the bankruptcy stage. In this regard, I wonder what is the real cost and human toll generated by this extraordinary display of corporate decadence. How many broken homes? How many foreclosures? How many suicides? Probably, we'll never know.

What it is apparently known, however, is that in 2010, according to a published press report, FSD spent more than 300 thousand dollars in personal jet travel (Reuters, 2011b). Arguably, that opulent lifestyle might have saved the jobs of a couple of pilots. Keep in mind that FSD was also part of the President's Council on Jobs and Competitiveness (Reuters, 2011b).

7.12 Back to research

From 1997 until 2000, I worked for John Pardo as part of a small team in charge of the design and construction of industrial size automated laser printers. Since these laser printers were used to expose photographic and digital paper they were known internally as "laser sensitometers." We ended up producing a line of highly-reliable voiced-commanded laser printers that were used in the quality control process of the paper produced by the company which by the late 1990s was still a major source of income. The last machine we built was commissioned to print in the nanosecond regime (Duarte *et al.*, 2005b). The astute business man who commissioned it, said that this high-speed industrial laser printer paid for itself soon after it was brought into operation. I must say that it was fun to work with honest, no-nonsense, engineers like Brian Reed. Also John Pardo, and his boss, were quite supportive of my interaction with a local university (SUNY) where I went to teach, twice a week, physics and astronomy.

In my new job, besides my work on laser printers, I quickly gained a reputation solving difficult optics problems and one of the lab heads would refer to me as his "top gun." Focusing on technical problems full time, was indeed a refreshing challenge.

Also in 1997, on return from my European tour, I made a detailed proposal to develop 3D imaging based on my interferometric laser techniques. This idea was supported by a few respected senior scientists and was given some consideration, however, it was not funded. Since this was not part of my research focus, I carried on with my laser optics and metrology projects.

In the early 1990s, my parents visited Rochester and I took my father to visit my lab and observe some experiments. That afternoon, as we were leaving he asked: "What's that smell, son?" He was referring to a pervasive low-level acidic smell, a smell we had become used to.

By the late 1990s my original observation of Eastman Kodak, as a chemical company, took a new emphasis. From time to time I would hear about people getting ill. Albeit this could have been within acceptable statistical limits, I was concerned. It was with the sudden death of a prominent spectroscopist, due to cancer, that this issue was highlighted even further. At the time, this well-known scientist was housed in a building next to mine. Although I had always been careful to stay away from chemistry and used in my laboratory only ethanol as a solvent, I was reminded of the danger every time I smelled something strange (see, for example, Orr (2007)). One night, I had a dream in which I was escaping with a gas mask on while most people around me were dead. I took some measures to ensure my safety. Thus, I kept a gas mask at hand to be used in case of fire or chemical spill. I also reminded myself it was time to leave.

By the end of 2000, a top manager in the Research Laboratories, a chemist that I used to refer to as Dr. No, decided to transfer Brian and I back to research. In essence, we were the last two experienced optical metrology scientists in residence. Amazingly, my laboratory was moved by the fourth time. In nearly twenty years a full geographical circle was completed. Brian was quite critical of our new boss, never moved to the research complex, and in 2004 decided to take an exit package. I contemplated this alternative but decided to stay in to reach my 20th year.

All of my original friends and colleagues were now gone. I could hear the echo of my own foot steps while walking through empty corridors. Bob James, my Australian co author and an excellent materials scientist, encouraged by the policies of Dr. No, also left.

Incidentally, during this period it was recommended that I should give a talk to some production and business managers on the new laser materials Bob and I had just discovered and patented (Duarte and James, 2003, 2004). They needed new products, I was told. As I began my talk, I was questioned by a particular manager about organizational issues, rather aggressively. Since the questions were not technical, and framed in an antagonistic tone, I declined to answer them and proceeded to leave the meeting. The talk did not even last one minute, the shortest talk of my career! I was appalled at the total absence of entrepreneurial energy among those businessmen. Phone calls and complaints quickly followed. Susan Parulski, my patent lawyer friend, said that my response was perfectly legal, particularly since I had avoided a confrontation (more on patents, later). Later, after returning from a trip, I learned that the poor manager that questioned me... had died of a heart attack.

Besides my experiments one of the few reasons to smile were some courses that I designed and gave to my colleagues. The subject matter was based on my sixth book title *Tunable Laser Optics* (Duarte, 2003) that besides lasers, focused on Dirac's approach to quantum mechanics. Among my students were some newly found friends and co authors: W. G. Ahearn, L. S. Liao, A. M. Miller, R. H. Nuttall, and K. M. Vaeth. Note: in addition to *Dye Laser Principles* (Duarte and Hillman, 1990), other book titles are *High Power Dye Lasers* (Duarte, 1991), *Selected Papers in Dye Lasers* (Duarte, 1992), *Tunable Lasers Handbook* (Duarte, 1995c), and *Tunable Laser Applications* (Duarte, 1995d).

In 2005, again it was time to contemplate taking the separation package. However, I was in the middle of an important experiment that would eventually yield the world's first demonstration of coherent emission from an electrically-excited organic semiconductor (Duarte *et al.*, 2005). Again, I decided to stay for one more year. True physicists have a weakness and a virtue. We live by our physics, from experiment to experiment, from new equation to new equation. The rest is secondary. When I first came to Eastman Kodak I came "for no more than five years." It was now just over 20.

One day, in the summer of 2005, hundreds of books were placed in giant boxes to be tossed out. Among them, numerous physics classics. All of them, with a pompous red stamp: DISCARD. Apparently, managerial minds needed office space and, in an increasingly empty company, elected to use the space in the book section of the beautiful research library. It felt like being in a corporate Gulag. In *one* of my last acts of defiance I wrote a critical letter to the CTO and cc the CEO (that is, FSD). No response.

Following the banishing of books I concluded there was no future for research, at the new transformed entity, which by now, with its limited vision, considered inkjet printing as the new frontier. To me, this meant little hope for a turnaround. This assessment was despite having personally developed, and demonstrated, an elegant interferometric technique, based on Duarte (1993a, 1993b), to measure the spatial resolution of inkjet and laser prints.

It was time to facilitate my exit. This resulted in my departure, with full separation benefits, due to yet another predictable downsizing, in April of 2006. Direct access to my separation benefits was made possible by a useful and pragmatic reform, adopted during the Fisher era, that allowed me, and others like me, not to participate in the standard company retirement scheme.

My last two papers under the banner of Eastman Kodak Company were: Duarte *et al.* (2006) that reported on a new green tunable laser and Duarte (2006) which reported on generalized dispersion equations for positive and negative multiple-prism refraction.

When we are born, the only certainty to follow is death. The time between these two events is our only capital. True success can be determined by how we control this time and by how we keep time thieves away. On April 2006, I became master of my own time.

Soon after my departure the demolition of buildings at Kodak Park began in earnest (Daneman, 2007).

7.13 Reflections

Late one glorious summer afternoon, in 2006, as I was flying back from the Finger Lakes region I observed the sun over a beautiful golden horizon. I was free, I could sense and live my freedom.

As I look back at my tenure at Eastman Kodak I see clearly two epochs. The first one, characterized by great products, technical challenges, and fulfillment as a scientist. Useful, knowledgeable managers, bright colleagues, helpful staff, and great friends. As a physicist, I had fun. It was a rich, creative, and unique experience. Indeed, I continue to be grateful to the magnificent legacy of George W. Eastman for having made that wonderful reality possible.

The second era, as the company went from Eastman Kodak to the new transformed entity, in my opinion, was a period highlighted by reduced transparency, reduced trust, increased entropy, confused vision, lost opportunities, waste, and instability. The one transformation sorely needed: a reduction in managerial ranks, was not observed. Whether this minimized and troubled entity survives in some form, or not, is irrelevant. Regardless, the sterling icon I was privileged to know… is no more.

Finally, I have no doubt that, despite the decline of film, all the troubles and tribulations of Kodak were unnecessary. This tragedy could have been avoided with a practical and clever leadership, driven by a belief in its people, utilitarian science, and technology.

References

Brayer, E. (1996). *George Eastman: A Biography.* The Johns Hopkins University, Baltimore.

Daneman, M. (2007). Kodak buildings set to be razed. *Democrat and Chronicle,* June 26.

Daneman, M. (2008). Kodak slips to third among local employers. *Democrat and Chronicle,* January 31.

Daneman, M. (2009). Kodak picture worsens. *Democrat and Chronicle,* May 1.

Daneman, M. (2012a). Kodak sales fall 27%; first-quarter loss grows to $366 million. *Democrat and Chronicle,* April 28.

Daneman, M. (2012b). Kodak announces 140 more local job cuts. *Democrat and Chronicle,* May 4.

Duarte, F. J. (1987a). Generalized multiple-prism dispersion theory for pulsed compression in ultrafast dye lasers, *Opt. Quantum Electron.* **19**, 223-229.

Duarte, F. J. (1987b). Pulsed dye lasers for spectroscopy and photochemistry. In *Kodak Laser Dyes* (Birge, R. R., ed.) Eastman Kodak, Rochester, pp. 6-9.

Duarte, F. J. (1989a). Optical device for rotating the polarization of a light beam. *US Patent* 4822150.

Duarte, F. J. (1989b). Ray transfer matrix analysis of multiple-prism dye laser oscillators. *Opt. Quantum Electron.* **21**, 47-54.

Duarte, F. J. (1989c). Pulsed dye lasers for spectroscopy and photochemistry. In *Kodak Optical Products* (Birge, R. R., ed.) Eastman Kodak, Rochester, pp. 6-9. [Note: this is an enlarged second edition to the first 1987 publication].

Duarte, F. J. (1990). Prismatic pulse compression: beam deviations and geometrical perturbations. *Opt. Quantum Electron.* **22**, 467-471.

Duarte, F. J. (ed.). (1991). *High Power Dye Lasers,* Springer, Berlin.

Duarte, F. J. (ed.). (1992). *Selected Papers in Dye Lasers,* SPIE, Bellingham.

Duarte, F. J. (1993a). On a generalized interference equation and interferometric measurements, *Opt. Commun.* **103**, 8-14.

Duarte, F. J. (1993b). Electro-optical interferometric microdensitometer system. *US Patent* 5255069.

Duarte, F. J. (1994). Solid-state multiple-prism grating dye-laser oscillators. *Appl. Opt.* **33**, 3857-3880.

Duarte, F. J. (1995a). Interferometric imaging. In *Tunable Laser Applications* (Duarte, F. J., ed.). Marcel Dekker, New York, Chapter 5.

Duarte, F. J. (1995b). Solid-sate dispersive dye laser oscillator: very compact cavity. *Opt. Commun.* **117**, 480-484.

Duarte, F. J. (ed.) (1995c). *Tunable Lasers Handbook*, Academic, New York.

Duarte, F. J. (ed.). (1995d). *Tunable Laser Applications*, Marcel Dekker, New York.

Duarte, F. J. (1997). Interference, diffraction, and refraction, via Dirac's notation. *Am. J. Phys.* **65**, 637-640.

Duarte, F. J. (1999). Multiple-prism grating solid-state dye laser oscillator: optimized architecture. *Appl. Opt.* **38**, 6347-6349.

Duarte, F. J. (2001). Laser sensitometer using multiple-prism beam expansion and polarizer. *US Patent* 6236461.

Duarte, F. J. (2002). Secure interferometric communications in free space. *Opt. Commun.* **205**, 313-319.

Duarte, F. J. (2003). *Tunable Laser Optics*, Elsevier-Academic, New York.

Duarte, F. J. (2005). Secure interferometric communications in free-space: enhanced sensitivity for propagation in the metre range. *J. Opt. A: Pure Appl. Opt.* **7**, 73-75.

Duarte, F. J. (2006). Multiple-prism dispersion equations for positive and negative refraction. *Appl. Phys. B* **82**, 35-38.

Duarte, F. J. and Hillman, L. W. (eds.). (1990). *Dye Laser Principles.* Academic, New York.

Duarte, F. J., and James, R. O. (2003). Tunable solid-state lasers incorporating dye-doped polymer nanoparticle gain media. *Opt. Lett.* **28**, 2088-2090.

Duarte, F. J., and James, R. O. (2004). Spatial structure of dye-doped polymer-nanoparticle laser media, *Appl. Opt.* 43, 4088-4090 (2004).

Duarte, F. J., Liao, L. S., and Vaeth K. M. (2005). Coherent characteristics of electrically excited tandem organic light-emitting diodes. *Opt. Lett.* **30**, 3072-3074.

Duarte, F. J., Liao, L. S., Vaeth, K. M, and Miller, A. M. (2006). Widely tunable green laser emission using the coumarin 545 tetramethyl dye as the gain medium. *J. Opt. A: Pure Appl. Opt.* **8**, 172-174.

Duarte, F. J., and Paine, D. J. (1989). Quantum mechanical description of *N*-slit interference phenomena. In *Proceedings of the International Conference on Lasers '88* (Sze, R. C., and Duarte, F. J., ed.). STS, McLean, VA, pp. 42-47.

Duarte, F. J., Reed, B. A., Burak, C. (2005b). Laser sensitometer. *US Patent* 6903824.

Feynman, R. P. (1985). *Surely You Are Joking Mr. Feynman.* W. W. Norton, New York.

Fisher, G. M. (1995a). Private communication.

Fisher, G. M. (1995b). Private communication.

Fisher, G. M. (1999). Private communication.

Google (2011a). http://www.google.com/finance?q=NYSE:EK &client=news

Google (2012a). http://www.google.com/finance?q=NYSE:EK &client=news

Google (2012b). http://www.google.com/finance?q=NYSE:EMN &client=news

Google (2012c). http://www.google.com/finance?q=PINK:FUJIY &client=news

James, T. H., *et al.* (1989). *Journey: 75 Years of Kodak Research.* Eastman Kodak, Rochester.

Kodak (1990a) *Laboratory Chemicals*, Eastman Kodak, Rochester.

Kodak (1990b) *Products for Polymer Science*, Eastman Kodak, Rochester.

Kodak (1990c) *Optical Products*, Eastman Kodak, Rochester.

Kodak (2012). Eastman Kodak Company and its U.S. subsidiaries commence voluntary chapter 11 business reorganization. *Eastman Kodak Company*, January, 19.

Hyde, W. L. (2006). Why 1916? A look back at OSA's roots. *Optics and Photonics News* **17** (1), 18-19.

Mattioli, D., and Fitzgerald, D. (2011). Kodak reshuffles deck. *The Wall Street Journal*, December, 23.

Mattioli, D. (2012). Kodak shutters camera business. *The Wall Street Journal*, February, 10.

Musashi, M. (1982). *A Book of Five Rings*. Overlook, New York [Note: this is a reprint of the original book written in 1645].

Orr, S. (2007). Pollution limits Kodak Park's uses. *Democrat and Chronicle*, August, 27.

Reuters (2011a). Eastman Kodak Company sells image sensor business to Platinum Equity. http://www.reuters.com/article/2011/11/08/us-eastmankodak-id USTRE7A7087 20111108

Reuters (2011b). As Kodak struggles, Eastman Chemical thrives. http://www.reuters.com/article/2011/12/24/us-eastman-kodak-idUSTRE7BN06B20111224

Reuters (2012). Photography pioneer Kodak files for bankruptcy. http://www. reuters.com/article/2012/01/19/us-Kodak-idUSTRE80I08G20120119

Schrödinger, E. (1992). *What is Life? With Mind and Matter and Autobiographical Sketches*, Cambridge University, London. pp. 115-116. [Note: this is a reprint of Schrödinger's early essays published in 1944 and 1958, respectively].

Spector, M., and Mattioli, D. (2011). Kodak seeks help as fears mount. *The Wall Street Journal*, October, 1.

Spector, M., and Mattioli, D. (2012). Kodak teeters on the brink. *The Wall Street Journal*, January, 5.

Tzu, S. (1996). *The Art of War*. Dove, West Hollywood [Note: this is a reprint of the original book written in the mid-fourth to mid-fifth century BCE].

Vanderbei, R. J. (2008). The earth is not flat. *Opt. Photonics News* **19** (11) 34-39.

Chapter 8

Observations I

8.1 Introduction
8.2 On the beauty of physics
8.3 On the status of physics
8.4 On my recent physics
8.5 On science and ethics
8.6 On the relevance of tunable lasers
8.7 On patents
8.8 On astronomy and extraterrestrial life
8.9 On the environment and climate change
8.10 On today's society
8.11 On economics
8.12 On modern medicine
8.13 On my father
8.14 On the future
References
8.15 Appendix I
8.16 Appendix II

8.1 Introduction

Motivated by my experiences, in academia and industry, I had the intention of writing a few pages on various observations on some aspects of physics, science, and ethics. Over time, however, this exercise resulted in a loose collection of opinions and short essays that reached beyond the original intent. Certainly, in the topics outside the realm of physics, my opinion should be considered just as the opinion of a concerned citizen.

8.2 On the beauty of physics

The concept of beauty is an essential part of life. For instance, beauty as defined by human symmetry and balanced proportions was a central theme in Ancient Greece, the cradle of Western civilization. Beauty in nature, as characterized by pristine mountains and transparent rivers, easily captivates and consumes the keen observer. And we all have our own preprogrammed concept of beauty when it comes to music. To live in the absence of beauty would be to remain submerged in the monotonous darkness of entropy.

Beauty in physics, is an abstract concept including a mathematical aesthetic quality, succinctness, and usefulness. Dirac, the foremost proponent of mathematical beauty in modern physics, once said "*it is more important to have beauty in one's equations than to have them fit experiment*" (Kragh, 1990). As much as I admire Dirac's physics, I tend to diverge a little with him on this one. Mathematical beauty looses its value if the equations don't agree with experiment. In this regard, beauty might be a necessary but not sufficient condition for good physics. Hence, one might say that in physics *it is the experiment that renders final judgment on the beauty of a theory*. In fairness to Dirac, it is said that he was also a proponent of the school of thought known as *instrumentalism* (Kragh, 1990) and according to this doctrine physical theories should relate to measurable quantities.

The definition of beauty and elegance in equations, as the definition of elegance and beauty in general, varies according to the observer. Besides the concept of aesthetics, succinctness is a very important element. Succinctness represents the power of a theory to describe nature in a compact, and useful, formulation. The lesser, the better.

More specifically, beauty in physics, as far as I am concerned, is exemplified by succinct useful mathematical expressions like Newton's gravitational law (Newton, 1726), Maxwell's equations (Feynman, 1965a), Planck's quantum energy equation (Feynman, 1965b), Heisenberg's uncertainty principle (Heisenberg, 1927), Schrödinger's equation (Feynman, 1965b), and the quantum principles via Dirac's notation (Dirac, 1947; Feynman, 1965b) ... just to name a few.

One of the nicest compliments that I have received, as a physicist, was in the mid 1990s when after giving a talk on the generalized multiple-prism grating dispersion theory (Duarte and Piper, 1982a; Duarte, 1990) a physicist in the audience (whose name was Warburton) commented: "Your equations look like well-written music."

A further aspect of beauty in physics, not usually considered, is beauty in experimental physics: the powerful sight of a visible laser beam as it propagates through space, the awesome and colorful impression of a light beam as it disperses,... the incredibly complex, and yet orderly, light pattern resulting from *N*-slit interference, which is succinctly represented by the generalized equation for *N*-slit interference (Duarte, 1991, 1993a).

Note: the 1st edition of Dirac's book was published in 1930. However, it was the 3rd edition, published in 1947, that introduced the *bra-ket notation* also known as *Dirac's notation.* This is explained by Dalitz (1987).

8.3 On the status of physics

Physics is in essence an experimental science. It is the experiment that relates the thought process to nature. *The experiment is the ethos of physics.* The description of the observations, via the wonderful language of mathematics, is the theory. Good theories can be used to provide a cohesive, and accurate, description of experimental results and they can predict new experimental results.

Even astronomy, and cosmology, can be considered as depending on experimental physics. Crucial to these two sciences are the large telescopes and digital solid state detectors both tools developed in the realm of experimental physics. And what's more, these tools are used to observe the greatest experiment of them all: the one unfolding in the cosmos.

In my opinion the greatest physicist to ever live has been *Sir Isaac Newton*. He gave us the principles of optics (Newton, 1704), the reflective telescope (Newton, 1704), Newtonian mechanics, classical gravitational theory (Newton, 1726), and calculus. He gave us the foundations of the present cathedral of physical knowledge. He was a superb theorist and a sublime experimentalist. He was a whole and complete *physicist*.

After Newtonian mechanics, Maxwell equations, and Einstein's general relativity, the greatest triumph of physics has been quantum mechanics. In fact, without Maxwell equations and quantum mechanics present day technology including lasers, computers, and the internet, could not have been possible. All the physics mentioned here was developed via a vigorous interaction between the human thought process and the experiment.

As I write today (2012) quantum mechanics, introduced in the mid 1920s, remains as the most successful theory of physics. The revolution initiated by quantum mechanics yielded technological breakthroughs in the form of the transistor (1954), the maser (1953), and its visible version, the laser in 1960. Remarkably, the quantum physics of Dirac, Heisenberg, and Schrödinger produced tangible practical devices within 30 years. Quantum mechanics plus electromagnetism provide the bases of the wonderful technologies we enjoy today.

Dirac said that in his time it was easy for "second grade physicists to do first grade physics" and today "is very difficult for first grade physicists to do second grade physics" (Dirac, 1975). I agree. No physics development, since the quantum revolution, rivals the brilliance or importance of quantum mechanics.

Despite a strong current of disapproval, which peaked in the 1980s, related to issues of interpretation no one has been able to successfully contradict quantum mechanics. Post quantum mechanics, the most important piece of physics is the Standard Model although it suffers from various limitations and shortcomings recognized and accepted even by some of its creators (Ward, 2004).

In this context, the rise of additional post-quantum theories, such as string theory, as branches of physics, represent an interesting development. Freedom of expression is central to physics. However, given the absence of experimental data to support a new theory it should be more appropriate to consider such theory as a part of mathematics rather than physics. In other words, until experimental evidence surfaces in support of unique predictions of a new theory that theory should remain in the mathematical realm. This is not a new, or solitary, notion as it has been advanced previously by others.

Physics to me is a very personal affair. Physics is like the ideal lover who is always there, always willing to give, always beautiful. Thus, I remain ambivalent about "big science." This approach means layers of administrators, very large budgets, and large numbers of participants. Not my cup of tea. Nevertheless, I observe the activities around large particle physics efforts with keen interest since these might help elucidate some important features of the Standard Model, such as the Higgs boson. At this stage it might be necessary to reiterate that all this physics has its genesis in quantum mechanics.

On the inexpensive, or less expensive, side of physics there is *light* and the significant advances in the application of lasers including new optical materials, cloaking, high precision optics, laser communications, laser cooling, laser guide stars, laser processing, laser isotope separation, laser medicine, laser microscopy, laser spectroscopy, nonlinear optics, and ultrashort pulse laser science. The interaction of optics with nano devices is also very exciting. Theoretically, all these phenomena can be described using semi-classical and quantum mechanical tools.

One final observation: mankind only has a limited vision of the universe conveyed to us by our marvelous and yet limited senses. As such, in my opinion, a final theory is a misguided quest. In this regard, as technology advances, and our measurements improve, and our understanding of the cosmos is enhanced, theoretical understanding will always improve. Personally, I find the concept of a "final theory," or a "theory of everything," rather limiting. The fun of discovery will most likely last as long as the human race continues.

8.4 On my recent physics

Since 2006 (my post Kodak era, at Interferometric Optics) I have been focused in the design of specialized interferometric and dispersive optics for commercial applications. I have also continued my laser and optics research. In particular I have published on coherent organic semiconductor emission from sub micrometer cavities or nanocavities (Duarte, 2007, 2008). I also had a lot of fun as visiting professor at the University of Santiago where I applied a polarization technique, that I used to control the intensity of lasers in laser printers (Duarte, 2001a, 2003), to independently vary the intensity of the lasers needed in laser cooling experiments (Olivares *et al.*, 2009).

An additional contribution was the second edition of *Tunable Laser Applications* (Duarte, 2009a) with emphasis in the medical application of lasers. Besides being the editor, I authored six chapters, and co-authored two chapters (with R. O. James and T. M. Shay), of a total of 15 chapters.

A most wonderful opportunity, which required long hours of concentrated work, was the possibility of extending the generalized multiple-prism dispersion theory to higher phase derivatives (Duarte, 2009b). I published the first derivative in 1982 (Duarte and Piper, 1982a) and five years later the second derivative (Duarte, 1987). For many years researchers working on femtosecond lasers made explicit the need of equations to calculate the third derivative. Now I had the space, and necessary silence, to work on this wonderful problem. Using a Newton style iterative approach I derived an exact, and elegant, mathematical framework to calculate higher derivatives (3rd, 4th, 5th...N^{th}) for the generalized multiple-prism dispersion... at will (Duarte, 2009b). These equations were published in their full explicit glory, *macho style* (borrowing an expression from Charlie Brau).

Further very satisfying physics, has been the work on large *N*-slit interferometers which so far has led to the demonstration of intra interferometric path lengths from 35 to 527meters (Duarte *et al.*, 2010, 2011). This work has been carried out in conjunction with colleagues of the US Army and Aviation Missile Command. The experiment with the 527 m intra interferometric path length was performed outdoors at night and involved installing the laser and *N*-slit optics on the top of a building while propagating the interferometric signals to the top of a small hill. To say the least, this was a lot of fun. Besides its practical significance for clear air turbulence detection, and propagation studies, this work has the potential to lead to some basic developments in the quantum-classical boundary and reports on the very first superposition of diffraction patterns on *N*-slit interferograms. This is relevant to "soft quantum measurements" and utilizes spider silk fibers in interferometry, which is another first. And there is more in the pipeline, the fun continues!

8.5 On science and ethics

There can be no science without ethics. Ethics is the framework that sustains science. My first introduction to ethics was provided by my maternal grandfather who could not tolerate corrupt behavior. My introduction to ethics in science was conveyed by listening and observing my teachers: R. A. Aitchison, R. E. B. Makinson, J. A. Piper, and J. C. Ward, among others. Although I received no formal training on the subject, I was privileged by a wealth of implicit and explicit teachings. I never questioned these teachings since they made intuitive sense to me.

The basics are the abstract concepts of honesty and truthfulness. Central to this ethos is to acknowledge the work and contribution of other physicists and authors. In this framework, the concept of plagiarism was, and is, completely alien. To me, personally, this framework was natural since from the beginning I had somehow come to consider my physics as an expression of my inner self. As such, I have never felt a desire to imitate my contemporaries.

Training as an experimentalist brings up also the issue of laboratory etiquette: never disturb someone else's experiment. Never take a piece of equipment without asking. The implication is clear: physics is a game of gentlemen, and ladies.

In recent times, it appears that there has been a deterioration in the observance of ethical guidelines. I have personally written several comments and letters on papers that have failed to cite my work. These range from the simple innocent omission to the publishing of some of my equations, duly disclosed in the open literature, as "new" by other authors. There is also the bizarre case of attributing to someone else some of my work (Duarte, 2006). In all, I have only published comments and letters in regard to the more blatant cases choosing to ignore others. In this regard, many victimized authors don't stand up in fear of alienating the system. Others, simply give up.

The issue of plagiarism appears also to be gaining ground and some prestigious refereed publications now clearly promote their code of ethics. Exaggerated language also distorts the meaning of publications with some authors trying to attach a grater importance to their reports. Worse is the case of inflated and exaggerated claims from marginal or non corroborating data.

In general, there is a need to begin to use the language properly again. Language should not be prostituted. When someone says "we stole this" he/she should be brought to order. They should be asked whether they are engaging in plagiarism. If not, they should say "we adopted this from author *x*" or "we did this after author *z*." The twisted, and incorrect, use of language can lead down a path of decadence.

On the issue of refereeing: anonymous peer refereeing, as a whole, has served science well. However, then again, it is based on honesty. Without honesty, the corner stone of ethics, the system breaks down. Example: a colleague submitted an important paper as a rapid communication. The referees sat and sat on the paper for months. Eventually, the paper was published but was soon followed by a paper, from a well-known institution, based on the same principle applied to an extended configuration. The second paper got most of the citations. In two occasions I've had to withdraw papers, and submit them elsewhere, because referees have sat on them and the editors have been either too busy or unwilling to get involved. The job of a referee is to review promptly, objectively, and to ensure that the work does include the relevant references. Often this can be a demanding task. The job of an editor is to select the appropriate referees, ensure balance and fair arbitration, all of this while expediting the process along. It is not easy.

In 2011 I wrote a comment to a paper pointing out a doubly erroneous equation plus the omission of references. After the comment was refereed, and accepted, the journal in question took the highly unusual decision to ignore the comment and to allow a "corrigendum" from the authors to correct their blatant mistake. To say the least, this is borderline unethical. I do not name the journal here because I love it. In this journal I published one of my first papers and later in my career I became a long serving Topical Editor. Moreover, I have a lot of respect for the founding editor. However, a brief Google search should reveal the details of this matter.

Up until recently, the physics community had been relatively free of cases of fraud. Over the years I had read about cases of fraud in science, but mainly outside the domain of physics. Sadly, this changed in 2000 due to a widely publicized case of apparent fraud that took place at Bell Laboratories. This work was published, via a series of papers, in the prestigious *Physical Review*, *Science*, and *Nature* magazines. Without revisiting the case, it is fair to suggest that although the system initially failed it eventually came into gear examining and marginalizing the work in question. It could be argued that this process of scrutiny was facilitated by the fact that the work in question was experimental. Errors and mistakes, via faulty data, misinterpretation of data, or a combination of these are part of the process. Sometimes these mistakes end up in print. When that happens the authors have the obligation to correct their own errors.

In addition to the topics mentioned there are wider and more extensive issues of ethics some of which can even be more subtle. John Ward, for instance, repeatedly complained about what he called "Ph.D. factories" in theoretical physics (Duarte, 2009c). He thought it was unethical to train and produce large numbers of new Ph.D.s in areas with notorious employment difficulties. Also, occasionally, he would pronounce the word "crook" in reference to certain professors leading these Ph.D. factories. Subtle hints of these criticisms are evident in his memoirs (Ward, 2004).

In one of my last conversations with Ward, he expressed skepticism on the value being attached to some research associated with large number of citations. In this regard, it should be indicated that some famous papers, with very large number of citations, in fact, have turned out to be erroneous. Ward was concerned at non-independent citations generated by groups of friends and students. Regardless of these doubts, it is superficial and even misleading to consider the raw number of citations of a particular paper in a given subfield, without considering the total number of publications in that particular subfield. In this regard, some scientists, bent on publicity, tend to exploit the ignorance of the media and the public to gain funding and notoriety.

To illustrate the point above let us consider a particular paper with 1427 citations. This is certainly a lot of citations when the number is considered in isolation. However, when the total number of publications in the area is identified as 330 000 then the situation changes completely since the ratio becomes ~ 0.43%. By comparison, in a different subfield, a paper is found to have 73 citations. This might look rather thin by itself but again the situation changes completely when the total number of publications in that area is found to be only 229 thus giving a ratio of ~ 32%. The first example corresponds to a theoretical area where a lot is published while the second example corresponds to an experimental subfield where a publication of a paper is preceded by measurements in an experiment that requires a laboratory infrastructure plus funding. Albeit these particular examples are obsolete by now, the principle applies.

An additional problem, associated with the present use of citations, originates from papers with large number of authors. These publications generate citations regardless of the weight of the contribution of a particular author. Thus, authors that play a peripheral role in a group can end up with more citations than sole authors that publish original research individually.

One further point: as demonstrated by the bibliometrics of Dirac, some good papers, can have a significant time lag from the time of publication to the time of citation (Kragh, 1990).

8.6 On the relevance of tunable lasers

During, and following, my doctoral research I did a lot of work on high-power copper-vapor-laser pumped dye lasers. In particular, we introduced narrow-linewidth multiple-prism grating oscillators (Duarte and Piper, 1982b, 1984). These are very high resolution , extremely low-noise, tunable high-power lasers. At the time this research was partly financed by the Australian Atomic Energy Commission. However, we also knew that the application of these lasers went beyond the nuclear field and into medicine in particular.

The papers that I published with Jim Piper were the first to describe the architecture and physics of copper-laser-pumped high-resolution pure multiple-prism grating laser oscillators and as such became seminal contributions to the field. A specific new oscillator architecture was the multiple-prism near grazing-incidence grating cavity. The interest was particularly keen among researchers working on laser development (Schäfer, 1990) and among scientists involved in uranium atomic vapor laser isotope separation (AVLIS) (Murray, 1987).

Prior to my departure from Australia (in 1983) I corresponded with Senator Sir John L. Carrick, at the time federal minister for energy, in an attempt to interest the Australian Government in AVLIS since Australia is one of the main producers of uranium. Following some correspondence he wrote: "The Uranium Enrichment Group of Australia (UEGA) assessed a number of competing technologies... and concluded that centrifuge technology would be best suited for an Australian plant... I am assured that UEGA has taken into account the recent advances in laser isotope separation... UEGA was not, however, prepared to commit to a technology still at the R&D stage" (Carrick, 1982). Obviously I did not succeed in this effort or as Australian nuclear and laser physicist Arthur Pryor would put it: "...the cautious hand refrained." (Pryor, 1997; Duarte, 2010). My premise was that nuclear energy did offer a viable alternative to fossil fuels and that the issue of safe waste disposal would eventually be solved using technologies such as crystallization (Ringwood, 1978).

As our papers were published I began to receive numerous postcards with reprint requests from the USA, all over Europe, Israel, and Asia. In particular I got several postcard reprint requests for our papers from colleagues at India's Bhabha Atomic Research Centre which was the center responsible for the nuclear tests of 1998. Among researchers -in the field- that acknowledged to apply our laser architectures were physicists from India (Singh *et al.*, 1994) and Japan (Sugiyama *et al.*, 1996). Today, the papers by Duarte and Piper (1981, 1982b, 1984) continue to be cited.

Besides laser isotope separation, our tunable laser oscillators have been widely applied by researchers working in atomic and molecular laser spectroscopy which are important subfields to both physics and chemistry (see, for example, Murray, 1987; Eckbreth , 1988; Sneddon *et al.*, 1996; Demtröder, 2008).

At the time that I was doing work, on high-resolution tunable lasers, Jim Piper was also working with some medical researchers to apply red lasers in the treatment of skin cancer which is very prevalent in Australia. The process is called photodynamic therapy and was originally pioneered in Buffalo, Western New York, by T. J. Dougherty. In this approach the patient is injected a dye which is then naturally cleansed out of the body by healthy tissue but retained by cancerous tissue. Illumination of the dye diffused into this tissue, by an appropriate laser, causes a photochemical reaction lat leads to the destruction of the cancerous cells (Goldman, 1990).

Jim Piper was using, however, a fixed wavelength laser. When I arrived at Eastman Kodak and became acquainted with the enormous number of dyes, available to Kodak researchers, I thought that it would be natural to start a project on laser photodynamic therapy. To this effect I did convince a lawyer to apply for a patent describing a dual laser system in which the pump laser would be used to detect the cancer and the optically-pumped tunable laser to eliminate it (Duarte, 1988). I also proceeded to talk to some business managers to secure some funding. Following a pleasant meeting I was told: "When you come up with a 300 dollar laser please comeback." I responded with a smile. Later, when the patent application hit some routine snag, it was abandoned. Fortunately, the idea was published in the European system and I've been pleased to learn that it has been useful to other researchers in the medical field. In 2005, a manager came back asking about dyes and photodynamic therapy. This topic had now become a business project. Again, I smiled. One aspect of this work that found commercial success was the development and demonstration of coumarin tetramethyl dyes as laser dyes (Chen *et al.* 1988; Duarte, 1989).

A perspective on the significance of our tunable lasers to cancer photodynamic therapy is given by Goldman (1990). Also, my work on generalized N-slit interferometry (Duarte, 1993a; Duarte, 1993b) has been useful to researchers in medical diagnostics (Sliski, 1997; Ortyn *et al.*, 2000; Kwok and Lee, 2005).

My contributions to multiple-prism theory (Duarte and Piper, 1982a; Duarte, 1987) have found applications in astronomy (Sirat *et al.*, 2005), laser microscopy (Nechay *et al.*, 1999; Siegner *et al.*, 2001), and laser pulse compression (see, for example, Osvay *et al.* 2005; Diels and Rudolph, 2006).

Perhaps the most immediately rewarding work I did, in a utilitarian sense, was the work on multi-color, voice-controlled, automated, large-scale laser printers which were called internally (at Eastman) "laser sensitometers" (Duarte, 2001a; Duarte *et al.*, 2005). This was true team work (mainly with Brian Reed and Chris Burak). As mentioned earlier several of these printers were built thus providing jobs and income for many.

The examples mentioned here illustrate the enormous breath of applicability of tunable laser physics. In between the examples cited we have a wide variety of scientific, industrial, and military applications. The second edition of *Tunable Laser Applications* (Duarte, 2009a) lists more than thirty application fields in its introduction.

Finally, it should be emphasized that physics developments, and resulting technology, can be put to work in a wide spectrum of applications and the physicist should not be paralyzed by this reality. In the case of doubtful uses it is not the physics that should be questioned. What should be questioned are the human intentions driving that application.

8.7 On patents

As a physicist I've always tended to publish in the open refereed literature as a first priority. However, sometimes, it is possible to publish and to patent (see, for example, Duarte 1993a, 1993b). Patents can be expensive. From a business perspective, patents make sense when there is a clear chance that the patent can lead to a product for manufacturing in a near or immediate future.

From an alternative perspective, I think that patents tend to be legal and priority documents that cover and protect mainly practical inventions and technological improvements with revenue generating potential. Hence, their purpose is to safeguard legal and economic interests. On the other hand, scientific publications are more general in scope as they can encompass technological inventions as well as being an avenue to disclose new science.

At Eastman, during the good times, there was a formidable legal department. A feature that was common in large US corporations. Under those circumstances, it was possible to freely assess the patentability of new ideas. There, I had the pleasure of working most of the time with Susan Parulski, although I also worked with a former Bell Labs attorney called E. P. Brandeau (Duarte, 1993b). All of my nine patents refer to unique and distinct ideas. This is not a strategy adopted by all inventors, some of whom tend to maximize the effect of an original idea by patenting numerous variations and permutations of the same idea. It is said that this approach is necessary to augment the protection of valuable product lines. On a related topic: monetary incentives to patent can result in an exaggerated number of patents and even in the patenting of frivolous ideas. Also, not all granted patents are valid.

As a physicist, I would always prefer to publish first, and patent second.

8.8 On astronomy and extraterrestrial life

As explained earlier, in 1997, I ended up in a new job in an optical and engineering laboratory associated with production. There, I began work on new industrial-size high-speed laser printers. I also continued my work on tunable lasers interacting quite a bit with outside colleagues. However, the immediate monotonous production environment began to bother me. I needed a break. In 1999, I linked up with one of the local State University of New York branches, the one at Brockport. The astronomy professor was retiring and they needed a replacement. I spoke to my boss, John Pardo, and he, very graciously, allowed me to teach on campus.

One afternoon, I met the old professor who was leaving partly disillusioned with the attitude of the students. He proceeded to list a litany of poor qualities that he saw in this new generation. One of the things that he said, and caught my attention, was that he was forced to teach astronomy "without mathematics." I was immediately intrigued. After all, mathematics is the language of physics and physics is all over astronomy. I kept the textbook *Explorations,* by Arny (1998) which offered a fairly good introduction to the subject. To the horror of some students the first two lectures were spent to give a most basic introduction to numbers, scientific notation, and yes, the meaning of simple equations. At the end of the first week of the first semester a bunch of students went to the head of the department and withdrew while expressing their disgust due to the use of mathematics. The poor chairman went into a panic mode, he mumbled something, but had no choice other than to accept the equations.

I taught general astronomy for two years (1999-2000). The first semester, there were some 60 students enrolled in the class. The last semester I taught the numbers had gone up to nearly 100. Perhaps the best way to describe what happened in this astronomy class is to quote a few of the anonymous comments written by the students:

• *At first I was scared but then I reacted...I actually enjoyed it!*

• *The course, and Professor Duarte, surprised me in that my eyes stayed open throughout the semester.*

• *You are not the everyday Professor, you know what you are talking about. You taught me things that I wanted to know and you made me want to come to class to learn more. You have a sense of humor, and have opened the doors of science to me.*

Some of the things I did in this class was to convince the students that traditional observational astronomy, which was covered in the experimental part of the course (by Matt Sidebottom), was only part of contemporary astronomy. The argument went on to say that in order to understand astronomy one had to become familiar with some basic techniques which included subjects such as mechanics, optics, and spectroscopy.

Thus, I taught them the basics parameters applicable to the design of telescopes and went on to introduce the uncertainty principle, and then, link this principle to the angular resolution of a telescope. We went on to discuss the advantages of space based telescopes and design parameters relevant to very large terrestrial telescopes. In a different part of the course a basic description of spectroscopy was linked with the Doppler effect and thus the concepts behind Hubble's law were introduced. Towards the end of the course we spent a bit of time on cosmology. In addition I would also discuss the latest, in the published literature, on subjects such as dark matter and dark energy. I would then emphasize that cosmology continues to be, a very exciting, rapidly evolving field where theoretical ideas are not always confirmed by observations. I would then introduce the second law of thermodynamics and explain to them that whichever the ultimate sequence of events turn out to be they would have to satisfy this all powerful law. This would be so unless an unknown truth, yet to be discovered, tells us otherwise.

I also engaged in some speculative thinking and back of the envelope calculations. Since everybody seems so interested about the possibility of finding extraterrestrial life, I spent a couple of lectures per semester on this subject. Given the enormous vastness of the cosmos, with estimates of over 10^{11} galaxies, it appears easy to reach a conclusion in favor of the existence of extraterrestrial life. Indeed, the search for extraterrestrial intelligence (SETI) has become a respectable academic pursuit. As a premise, it is reasonable to accept the existence of extraterrestrial life. However, my guess is that this life is most likely in bacterial forms and morphologies yet unknown to the human experience.

To estimate the likely probability of finding *human-like* extraterrestrial life, in our own galaxy, I engaged in a simple approach using published astronomical figures while ignoring temporal issues. The approach adopted was based on the usual multiplication of probabilities. However, the question that I posed was quite specific: "What is the rough probability of finding a planet *almost exactly like Earth* somewhere else in our galaxy?" Asking that rather restrictive question means that, as an initial step, a series of conditions have to be met in order to determine the likelihood of finding a star with the same mass and luminosity as our Sun. Also, this star would have to be positioned approximately at the same radial distance on the disk, as our Sun, from the center of the galaxy (a very restrictive condition). That is, it would have to be positioned on a ring with a thickness about twice the diameter of the solar system. Then, that G-class star would have to have a planetary system. Next, our candidate planet would have to have similar mass and volume, as Earth, and would have to orbit within the habitable zone, around the Sun-like-star. This rough back-of-the-envelope exercise yields a number of possible candidates of about 100. Certainly, relaxing the conditions imposed earlier leads to larger numbers. Also, it would be interesting to repeat this particular calculation using the latest available data in a formal context.

A crucial element, not included in the previous discussion, is the fact that the Moon plays an essential role in regulating the climate on Earth. We also know that the Moon ended up there following a very unique cataclysmic event. What is the probability for that event? That decreases the odds even further.

As discouraging as this seems it gets even worse if you extend the question to consider *humanoid life forms*. Thanks to the genius of physicist Luis W. Alvarez (1987), we know that the Earth was struck by a giant asteroid some sixty five million years ago. It was that collision, that wiped out the dinosaurs, it was that collision that eventually gave us a chance. Without that miraculous collision the Earth would belong the reptiles. Finding a planet like ours won't be easy, finding extraterrestrial life like ours, will be even harder.
Since these lectures were presented (1999-2000) a rare Earth hypothesis was highlighted in a book by Ward and Brownlee (2000).

Thus, it can be argued that our galaxy might not be teeming with human life forms albeit this argument does not exclude other life forms. Further, it can be argued that even if planets like our Earth are found out there they might well be populated by alternative life forms... like ferocious alligators, for instance. Nevertheless, given the total number of galaxies ($\sim 10^{11}$) there is a distinct probability of humanoid species out there, albeit separated by enormous and increasing distances. Although I made it clear that all this was just a conjecture, the students kind of liked it. Maybe it gave them a sense of our own uniqueness.

My interest in astronomy began in the mid to late 1990s when I began receiving questions from physicists working for astronomers about high-power tunable lasers. For ages poets have written about the twinkling of stars. This poetic twinkling is a nuisance, imposed by atmospheric turbulence, of particular relevance to very large terrestrial telescopes. One way to overcome this problem consists in the clever use of *adaptive optics* in conjunction with *laser guide stars*. As it turns out, at an altitude of about 90 km there is a layer of sodium which is about 10 km thick. This sodium, originating from meteors entering Earth's atmosphere, has a resonant transition at about 590 nm. This wavelength is near the maximum emission of high-power dye lasers using rhodamine 6G dye. When the sodium is illuminated at this wavelength it produces an emission. This emission is what is called a laser guide star. The information provided by the light from that transition, following propagation through the atmosphere, can then be used to deform, in real time, the optics of the telescope and thus compensate for the turbulence. Further, the lasers need to be narrow-linewidth and very low beam divergence which are two of the characteristics of the high-power narrow-linewidth oscillator amplifier systems that I had developed for spectroscopy and laser isotope separation. In this regard, it is important to configure narrow-linewidth laser oscillators yielding a beam divergence close to what is allowed by Heisenberg's uncertainty principle (Duarte, 1999, 2001b). When a laser does that, it is said to be "diffraction limited." Parenthetically, one of the first experiments in laser guide stars was performed using flashlamp-pumped dye lasers by Everett (1989) and colleagues (also see, Primmerman *et al.* (1991)). Today, additional laser alternatives, such as crystalline solid state lasers and fiber lasers, are also available for this application.

In addition to astronomy, I also taught a couple of courses in quantum mechanics based on the *Feynman Lectures on Physics*. One was an introductory course and the other a detailed course with emphasis on problem solving. Going back to the criticisms that the old professor formulated towards the student population: I noticed some differences relative to the times when I was a student and relative to the times I taught at Alabama in the mid 1980s. I was certainly not impressed by the lack of effort demonstrated by some. I was struck by smelling alcohol on a few students coming to class. I also became concerned to hear about some of the biological challenges faced by the present generation. However, at the same time I was inspired to see smiling young faces with bright open eyes eager to learn and eager to understand the wonders of nature.

8.9 On the environment and climate change

We all know that the "world" will eventually end. This is a result of what we have learned from physics and astronomy. We know that our beautiful star, the Sun, is about 5 thousand million years old. It has run about half its life cycle. He is like 50 going on to 100. Eventually, the Sun will grow into a red giant and might engulf the Earth in a deadly expansion prior to becoming a white dwarf. For the remains of the solar system, extremely low temperatures will follow after the death of the Sun. However, more recently, astronomers have suggested that the scenario just described might be surpassed by other cataclysmic events involving our galaxy.

Therefore, according to current knowledge, the long term picture is no good and there is nothing we can do about it, except prepare to eventually leave for another suitable younger planetary system. As we saw earlier, finding such planet won't be easy and going there, even if we had the means, will be rather difficult. Here, we should be conscious that according to Einstein's special relativity, and Lorentz's transformations, displacement at very high speeds, even a fraction of c, means that the time contraction experienced by the travelers (relative to the time on the Earth) will make any post-departure contact with Earth meaningless. However, we should not give up on new discoveries that might make possible a new age of exploration for cosmic navigators.

Back to Earth: climate is not a constant it has always been changing. For example, 65 million years ago, during the time of the dinosaurs, temperatures were much warmer than now. Even the Antarctic continent was warmer (Nova, 2011). Climate cycles come and go.

Given the long cold winters, in Western New York, from a selfish perspective, a warming trend would make life here a little bit easier. In fact, shorter winters would be a welcome change for the local wine industry and agriculture in general. However, there is no question that mankind has an enormous capacity to pollute and to ruin clean natural surroundings.

As a matter of principle, as far as the environment is concerned, the issue goes far beyond CO_2 or fossil fuel emissions. The main enemy is entropy. For the physical carbon based life we know, entropy is death. Almost everything we do, every process we engage in, every action we take, leads to an increase in entropy. One of the creators of quantum mechanics in 1946 referred to life as *negative entropy* (Schrödinger, 1992). This means that before we do anything, in a large scale, we must think very carefully and we must ask: is it really necessary? Do we need that new suburb? Do we need that new infrastructure? In this regard, mining of non-renewable resources (such as aluminum, copper, and gold) is particularly damaging to the environment and should only be conducted under well thought-out guidelines. Industrial predatory fishing is also very harmful and needs to be phased out. Economic slow-downs, recessions, and even depressions, help to keep entropy down. Unfortunately, the human cost of recessions can be high for those affected. More on this, later.

As far as human activity goes, it should be obvious that everything we do contributes to an increase in entropy. Many accepted human activities contribute to entropy, principal among them is economic growth based on increased consumption of raw materials. In this regard, globalization also contributes to an increase in entropy. But this goes unnoticed by most. Paradoxically, many in western societies appear to conveniently ignore or oppose nuclear energy, as alternative to the consumption of fossil fuels. Until large scale fusion arrives, nuclear energy should not be ignored, on the contrary, it should be encouraged.

I don't use air conditioning in the summer. Instead, I rely on the shadows of trees close to my house. In winter, I don't use central heating and the temperature inside the house can drop to about 5 C (close to 42 F) when the winds outside reach - 30 C (about - 22 F).

When outside temperatures really take a dive the temperature inside can be just above freezing. My electrical *and heating* bill for the month of February 2007 (the third snowiest February in record, in Western New York) was 46.96 US dollars. In January 2012 the bill went up to 85.21 US dollars, due to a number of factors. My approach to deal with low temperatures is the old fashion way of wearing appropriate clothing. I also have a very efficient cast iron stove where I burn wood collected from downed trees in the area. Wood burning is a natural process and in my view it is much better to dispose of dead wood in this manner rather than in landfills. Of course wood cutting, and axing, is hard work that requires good physical shape and attention to safety details.

Recently, my physicist friend Sergey Popov reminded me that in 1995 (at a laser conference in Charleston, South Caroline) I conjectured that it would be difficult, if not impossible, to give an affluent American standard of living to the vast populations of the former communist nations without affecting the environment for the worse. At that time, major corporations in the West were looking at China and other places as "new frontiers" for investing.

As bad as global warming appears to be, it pales into insignificance relative to a possible *global cooling*. A cooling of the Earth would mean certain and expedient death for most if not all of human and animal life as we know it. In this regard, the possible return of a glacial phase (as suggested by Fred Hoyle, in 1999), mega volcanoes, and corresponding volcanic winters, pose the greatest threat to human existence. And of course, there always the threat of the odd asteroid, as discovered by Luis Alvarez (1983).

Recently, I learned that Newton predicted "the end" by the year 2060 (Nova, 2007). Certainly, our wonderful planet will be here a lot longer after that, however, given the inexorable increase in entropy, surrounding large centers of population, present civilization, as we know it, may not.

8.10 On today's society

As a premise I will go back to some of Schrödinger's thoughts. In his writings of 1946 on ethics, he describes how ethics is necessary to control the primitive will and primitive appetites. Later he warns against excessive desires and that we are in "grave danger of missing the path to perfection" (Schrödinger, 1992). Building on his ideas, it appears that there is a human tendency towards decadence analogous to the natural increase in entropy, and cellular disorder, which culminates with the death of the human body. As entropy is a physical phenomenon that leads to disorder it is not difficult to imagine a manifestation of entropy in a social context.

In a free and open society, there must be personal responsibilities and self restraints not to degenerate in libertinism, chaos, and anarchy. And Schrödinger stipulates that the necessary controls are ethical principles. In my opinion the extraordinary freedom available in the US is the engine that generates scientific innovation and the creation of wealth. However, if abused, this freedom can lead to anarchy and anarchy is the ultimate form of societal entropy. Thus, freedom must be protected with a framework of ethical principles.

Observing today's societal trends it is not difficult to realize that there is a tendency to promote, and even commercially exploit, the human weakness toward primitive appetites. Vulgarity, obscenity, drugs, decadence, and violence, are some of the elements glorified by the popular culture. It appears that the old ethical constraints have been lifted and that personal freedoms are increasingly abused thus yielding libertinism and a societal drift towards anarchy.

An ethical frame of reference is also essential in the business world. The 2008-2009 debacle, in the American economy, is another example of the abandonment of ethics. The system was abused by the unethical above, the unethical in the middle, and the unethical below. Capitalism needs an ethical framework to work properly. The fear of failure, as a mean to impose discipline, should not be diluted by legislation. Without an ethical structure, and the option of failure, monopolies and corruption can drown free enterprise and creativity. Right now, we don't have a true free enterprise system, what we have instead is pseudo capitalism dominated by monopolies.

8.11 On economics

Since childhood days I wondered on what made people work. Later I became familiar with the concept of natural resources, the value of gold, inflation, etc. From Australia I observed hyper inflation and economic collapse in socialist Chile (1971-1973). After becoming familiar with working economies as in Australia and the US, I began to develop a basic understanding of the parameters that affect the economy. Mathematical models only apply to working economies.

The economy is intimately linked to the human will. Anything that destroys this *human will* also destroys the economy. A set of elements that can subdue the human will, include phenomena such as corruption, tyranny, and high taxes. Corruption strengthens monopolies, obliterates competition, and threatens democracy.

Corruption coupled with high taxes kill economies. In many countries, high taxes are the mechanism necessary to feed waste and the corrupt appetite of the controlling bureaucracy. Able and creative people do not want to participate in a system that will take away the fruits of their labor. That's why achievers from corrupt countries tend to emigrate. The tax code should be extremely simple and transparent. It should be neutral. It should neither help nor discriminate against particular sectors of society.

Notice that the above should not be interpreted as an advocacy for a system devoid of rules and regulations. What is needed are ethical and transparent laws for protecting life, dignity, and the fruits of individual labor, without suffocating creativity and innovation.

As hinted in a previous section, a further complication has to do with the issue of economic growth. The economies of the world, and the population of the world, cannot grow indefinitely fueled by the availability or raw materials since these materials are finite. In other words, the finite dimensions of our planet indicate that "old fashion" growth cannot be indefinite. Thus, a new model capable of motivating the human will of future generations will have to be found. My first guess is that we'll either have to go into space exploration and/or completely improve the human behavior and perspective, that is, find the "path to perfection" as hinted by Schrödinger (1992). My second guess is that space exploration will be easier.

Footnote: I should add that I tried to sell my Kodak shares in 1997 but was not allowed since those had been given to me when hired to be sold on exiting the company. Thus, I sold them in 2006 at nearly 26 dollars per share (early in 2012 those shares were less than 30 cents). Observing the stock market, and in particular the inflated stocks of a few poor corporate performers, I came to the conclusion, by December 2007, that the market was overvalued and entering a saturation phase. Hence, I decided to pull out of the market thus escaping the grief that came afterwards. It is amazing the pain you can avoid if you are a little bit observant. Also, avoid investing on financial products devoid of transparency.

8.12 On modern medicine

The western public looks at modern medicine, and its practitioners, in excessive reverence. Little do they know that many of the technical marvels in this area are due to advances in physics and engineering. That is, physics and engineering applied to medicine. In this exercise, the physicists and engineers that have made possible these advances get little or no credit. To illustrate this point the reader should think for a moment of a medical profession devoid of optics, lasers, MRI facilities, or advanced electronics.

Perhaps this irreverent attitude towards modern medicine was inherited from my teachers at Macquarie. Dick Makinson, for instance, told us that while at the University of Sydney, he would always check the physics grade of a physician before deciding to pay them a visit. Ron Aitchison, while teaching us how to design and build electronics for electrocardiograms, told us of the many poor patients that were killed in the early days due to faulty circuitry.

Albeit interested in biological and medical applications of lasers I have been, in general, fortunate to keep myself away from the reaches of "organized medicine." I have been briefly hospitalized only twice: following my second year at Macquarie (early 1975), due to appendicitis, and in the early 1990s due to a sports injury. The last time I had antibiotics was in the late 1970s and the only "medicine" I take is a small aspirin before shoving snow which can be physically demanding when coupled with very low temperatures.

My appendix was removed by a Roman physician called Roberto Robi who was the doctor of my brother's soccer team and more of a friend than just a medical practitioner. He practiced at a small private hospital in the Sydney suburb of Lewisham. Due to some extended initial inflammation I ended up staying in the hospital, with first class service, for a few days. I used this time to read the long version of *Don Quijote* and to befriend some of the staff. I remember particularly well an attractive classy young nurse who would come to my room to have afternoon tea. These events took place exactly at four o'clock with fine tea "with milk" and "biscuits" that she would bring. Thus, I became familiar with the operating theater of the small hospital and was educated about some of the usual mistakes during surgery. In this regard I was advised "to avoid operations on a Monday morning."

Not all my encounters with female medical practitioners have been as delightful as the Lewisham experience. When I first got to Kodak I was sent to the company's clinic to have a check up and to take a routine photograph of my retina, given my work with lasers. The examining physician was a female in her forties. Before I knew it, she was squeezing my testicles. Following an interrogating glance she said: "I had to make sure you didn't have testicular cancer." In 2006, I went for a physical and was examined by another female practitioner. Again, an exhaustive testicular examination. This time "to make sure there was no hernia."

In the mid 1990s I noticed that small area of dark skin was slowly growing in diameter on my right tight. I went to the physician and asked him to remove it. He refused arguing that it was cosmetic and the insurance would not pay. I got myself a new scalpel and made a nice square cut around the offending area and made it 4-5 mm deep. Following some initial bleeding it healed beautifully.

Maintaining distance from the medical profession requires work. I do so by consuming only the necessary calories and exercising daily. I don't eat red (or processed) meat and keep away from sugar, fat, and fast food. I weight about 73 kg (~ 161 lb, at a height of 179 cm or ~ 70") which is ~ 10% over the weight I had when at Macquarie.

Lasers are used in a plethora of medical applications including angioplasty, biostimulation, cancer diagnostic, cancer photodynamic therapy, cardiology, cosmetics, dentistry, dermatology, lithotripsy, ophthalmology, surgery, and urology. For a current discussion on the subject of laser medical applications see *Tunable Laser Applications* (Duarte, 2009a).

Many of today's society health problems are associated with bad societal habits and the expectation is that organized medicine should cure it all. Thus people over indulge, over consume, engage in risky behavior, and then expect the taxpayer to take care of the mess. Add to this some widespread tactical problems, such as poor diagnoses, unnecessary surgery, erroneous surgery, over use of antibiotics, rampant hospital infections, and a horrendous reality emerges.

In regard to unnecessary surgery: it is my opinion that some medical practitioners today don't even take seriously the Hippocratic Oath ("never do harm") anymore and prefer to promote risky surgeries over safer less costly alternatives.

Additional problems include the nasty side effects associated with many of the chemistry-based pharmaceuticals, the so called "medicines." Also, treatments such as cancer chemotherapy are borderline barbaric. One should also question the inability of the medical profession to come up with cures to wide spread diseases such as diabetes and the common cold.

Given the high cost of modern medicine, and the increasing numbers requiring care, western societies face a daunting dilemma. In this regard, people should be encouraged to take responsibility of their own health. Governments should be marginalized from decisions that belong to individuals and their immediate families. The reality is that those individuals that are better informed have an edge by taking responsibility of their own health and future. And, if it becomes imperative to interact with the medical profession... be well informed.

Finally, those who promise a future cure for "old age" conveniently ignore the second law of thermodynamics, conveniently ignore entropy. Albeit the process of ageing will be slowed down and the human life span increased, in the physical reality we know, there is no escape from the ultimate truth: human life, is a terminal condition. Not even stars, or galaxies, are eternal.

8.13 On my father

The tall handsome man, who could solve intricate problems, and work relentlessly on project after project, eventually grew old. Old age for Luis Enrique Duarte began to take its told when he was around 85, although, you could not tell that from his firm handshake.

Things got difficult in Chile for my parents and they decided to go back to Australia early in 2010, when he was approaching 89. Both my parents endured the long trans-Pacific flight from Santiago to Sydney on their own feet, without once requesting assistance.

In Sydney, my father continued his long daily walks until July, 2011, when he suffered an accident and ended up in a hospital with a broken right femur. I flew there in September, stayed for a month and returned to the US a week after he was allowed to go back home. Going home was a significant improvement. I went back again in November. My mission was to take him out, so that he could see the sky again and feel the sunlight. So he did, sporadically, as permitted by an unusual rainy spring.

My father knew he was facing death. Yet, he never complained and he would smile every time I would go to the side of his automated bed. The morning of the 21st of November he woke up, looked at me, smiled, and softly said: "Hola." He was already very weak since he had ceased eating, for a few days now. Water melon juice, and lemonade with honey, is all he accepted. He was neither in pain, nor was he being medicated.

Around noon we began to listen to some classical music, including Verdi's Va Pensiero (Chorus of the Hebrew Slaves), and the immortal Ave Maria. He was ready. At 3 pm, the man I thought invincible, departed this word. I had the privilege to hold his head at that all important and unique moment. It was a clean, honorable, and peaceful transition. That kind, gentle "Hello" was his last word.

Immediately following his departure we gathered around him, read Psalm 23, and gave thanks for his life. When the physician arrived to issue his death certificate we had him already perfectly dressed in his finest, and favorite, pure wool suit. Exactly three days later, on a rainy Thursday (24th) at 3 pm, according to Christian tradition, we laid him to rest at a little country cemetery, in Moss Vale, New South Wales. Mother cried, quietly.

Born in 1921, to Luis Duarte and Maria Diaz, grew up in a little country town during the depression. At age 10 he held two morning jobs. His first job, very early, at the local bakery, and the other, at the railway station. In the afternoon, he went to school. Eventually, he became a hunter, a mountaineer, a navigator, a self taught engineer, and entrepreneur. Married in 1951. Led his family to Australia in 1971, and died in his adopted country, in 2011. Albeit we lived at a distance for the last 28 years, not a day went by that I didn't think of him. Often we spoke on the phone, and I visited whenever possible. Now, I think of him, even more. The death of my father, brought me precious moments of unusual contemplation. That was his final gift.

8.14 On the future

Here, I provide a wish list of what I would like to see in the future. Of course, this is a much easier exercise than trying to predict the future. This list goes roughly from the more obvious to the less obvious:

1. A wireless world dominated by secure free-space electromagnetic and/or narrow-linewidth laser communications.
2. Three dimensional image capture (3D cameras). Laser projection of high quality 3D images (either interferometric or holographic) from cell phones, cameras, tablets, etc (a bit like in science fiction).
3. Accurate computerized wireless medical diagnoses. Compact electro-optical sensors to measure and transmit basic biomedical parameters (similar to those long advocated by science fiction).
4. Sale of clean and healthy fast food.
5. Bloodless surgery, and bloodless amputations, via the correct, widespread, standard, and skillful, use of lasers.

6. Efficient repair of broken bones without the use of metal rods, metal plates, or screws.
7. Wide spread and standard use of quantum computational techniques by engineers: a new breed of quantum engineers.
8. Mach 10 air travel: LA - Sydney in about 2 hours, lets say.
9. The elimination of corruption.
10. Side-effect-free medications.
11. Oral medications to prevent, and cure, dementia.
12. A cure for type I diabetes, or at least, insulin in oral form.
13. Design and creation of microscopic, or nanoscopic, life forms that would feed on cancer cells and simply die, or exit the body, once cancer cells are depleted.
14. Description of all measurable physical phenomena via quantum mechanics.
15. Directed energy, and kinetic energy, weapons to protect the planet from possible collision course asteroids (first hinted during SDI days?).
16. Complete theoretical understanding of the weather. Accurate, and reliable, predictions of climatic conditions.
17. Control of the weather: induce rain in deserts; dissipate storms.
18. Space travel at luminal or superluminal speeds (long advocated by science fiction).
19. Learn to control, and even defeat, entropy.

References

Alvarez, L. W. (1987). Mass extinctions caused by large bodily impacts. *Physics Today* **41** (7), 24-33.

Arny, T. T. (1998). *Explorations: An Introduction to Astronomy.* McGraw-Hill, Boston.

Carrick, J. L. (1982). Private communication.

Chen, C. H., Fox, J. L., Duarte, F. J., and Ehrlich, J. J. (1988). Lasing characteristics of new coumarin-analog dyes: broadband and narrow-linewidth performance. *Appl. Opt.* **27**, 443-445.

Dalitz, R. H. (1987). Another side to Paul Dirac. In *Paul Adrien Maurice Dirac* (Kursunoglu, B. N., and Wigner, E. P., eds.). Cambridge University, Cambridge. Chapter 10.

Demtröder, W. (2008). *Laser Spectroscopy: Basic Principles*, 4th Ed., Springer, Berlin.

Diels, J. C., and Rudolph, W. (2006). *Ultrashort Laser Pulse Phenomena*, 2nd Ed., Academic, New York.

Dirac, P. A. M. (1947). *The Principles of Quantum Mechanics*. 3rd Ed., Oxford, London.

Dirac, P. A. M. (1975). Lecture delivered at the University of New South Wales, Sydney, Australia.

Duarte, F. J. (1987). Generalized multiple-prism dispersion theory for pulse compression in ultrafast dye lasers. *Opt. Quantum Electron.* **19**, 223-229.

Duarte, F. J. (1988). Two-laser therapy and diagnostics device. *European Patent Publication* 0284330.

Duarte, F. J. (1989). Ray transfer matrix analysis of multiple-prism dye laser oscillators. *Opt. Quantum Electron.* **21**, 47-54.

Duarte, F. J. (1990). Narrow-linewidth pulsed dye laser oscillators. In *Dye Laser Principles* (Duarte F. J. and Hillman, L. W., eds.). Academic, New York, Chapter 4.

Duarte, F. J. (1991). Dispersive dye lasers. In *High Power Dye Lasers* (Duarte, F. J., ed.). Springer-Verlag, Berlin, Chapter 2.

Duarte, F. J. (1993a). On a generalized interference equation and interferometric measurements. *Opt. Commun.* **103**, 8-14.

Duarte, F. J. (1993b). Electro-optical interferometric microdensitometer system. *US Patent* 5255069.

Duarte, F. J. (1999). Multiple-prism grating solid-state dye laser oscillator: optimized architecture. *Appl. Opt.* **38**, 6347-6349.

Duarte, F. J. (2001a). Laser sensitometer using multiple-prism beam expansion and a polarizer. *US Patent* 6236461.

Duarte, F. J. (2001b). Multiple-return-pass beam divergence and the linewidth equation. *Appl. Opt.* **40**, 3038 - 3041.

Duarte, F. J. (2003). *Tunable Laser Optics*. Elsevier-Academic, New York.

Duarte, F. J. (2006). Equations for multiple-prism dispersion were derived earlier. *Laser Focus World* 42 (9), 11.

Duarte, F. J. (2007). Coherent electrically-excited organic semiconductors: visibility of interferograms and emission linewidth. *Opt. Lett.* **32**, 412-414.

Duarte, F. J. (2008). Coherent electrically excited organic semiconductors: coherent or laser emission? *Appl. Phys. B* **90**, 101-108.

Duarte, F. J. (Ed.) (2009a). Tunable Laser Applications. 2nd Ed., CRC, New York.

Duarte, F. J. (2009b). Generalized multiple-prism dispersion theory for laser pulse compression: higher order phase derivatives, *Appl. Phys. B* 96, 809-814.

Duarte, F. J. (2009c). The man behind an identity in quantum electrodynamics, *Australian Phys.* **46** (6), 171-175.

Duarte, F. J. (2010). Atomic vapor laser isotope separation: theAustralian contribution. *Australian Phys.* **47** (2), 38-40.

Duarte, F. J., and Piper, J. A. (1981). A prism preexpanded grazing incidence pulsed dye laser. *Appl. Opt.* **20**, 2113-2116.

Duarte, F. J., and Piper, J. A. (1982a). Dispersion theory of multiple-prism beam expander for pulsed dye lasers. *Opt. Commun.* **43**, 303-307.

Duarte, F. J., and Piper, J. A. (1982b). Comparison of prism preexpanded and grazing incidence gratings cavities for copper laser pumped dye lasers. *Appl. Opt.* **21**, 2782-2786.

Duarte, F. J., and Piper, J. A. (1984). Narrow linewidth high prf copper laser-pumped dye-laser oscillators. *Appl. Opt.* **23**, 1391-1394 (1984).

Duarte, F. J., Reed, B. A., and Burak, C. J. (2005). Laser sensitometer. *US Patent* 6903824.

Duarte, F. J., Taylor, T. S., Clark, A. B., and Davenport, W. E. (2010). The N-slit interferometer: an extended configuration, *J. Opt.* 12, 015705.

Dupre, P. (1987). Quasiunimodal tunable pulsed dye laser at 440nm: theoretical development for using a quad prism beam expander and one or two gratings in a pulsed dye laser oscillator cavity. *Appl. Opt.* **26**, 860-871.

Eckbreth, A. C. (1988). *Laser Diagnostics for Combustion Temperature and Species*. Abacus, Cambridge.

Everett, P. N (1989), 300 Watt dye laser for field experimental site. In *Proceedings of the International Conference on Lasers '88*. (Sze, R. C., and Duarte, F. J., eds.). STS Press, McLean, Va., pp. 404-409.

Feynman, R. P., Leighton, R. B., and Sands, M. (1965a). *The Feynman Lectures on Physics*, Vol. II. Addison-Wesley, Reading.

Feynman, R. P., Leighton, R. B., and Sands, M. (1965b). *The Feynman Lectures on Physics*, Vol. III. Addison-Wesley, Reading.

Goldman, L (1990). Dye lasers in medicine. In *Dye Laser Principles* (Duarte F. J. and Hillman L. W., eds.). Academic, New York, Chapter 10.

Heisenberg, W (1927). Über den anschaulichen inhalt der quantentheoretischen kinematik und mechanic. *Z. Phys.* **43**, 172-198.

Kragh, H. S. (1990). *Dirac: A Scientific Biography*. Cambridge, London.

Kwok, C. S., and Lee, K. Y. (2005). Microdensitometer system with micrometer resolution for reading radiochromic films. *US Patent* 6927859.

Murray, J. R. (1987). In *Laser Spectroscopy and its Applications* (Radziemski, L. J., Solarz, R. W., and Paisner, J. A., eds.). Marcel Dekker, New York, Chapter 2.

Nechay, B. A., Siegner, U., Achermann, M., Bielefeldt, H., and Keller, U. (1999). Femtosecond pump-probe near-field optical microscopy. *Rev. Sci. Instrum.* **70**, 2758-2764.

Newton, I. (1704). *Opticks*. Royal Society, London.
Newton, I (1726). *Principia Mathematica*. Royal Society, London.
Olivares, I. E., Cuadra, J. A., Aguilar, F. A., Aguirre Gómez, J. G., and Duarte, F. J. (2009). Optical method using rotating Glan-Thompson polarizers to independently vary the power of the excitation and repumping lasers in laser cooling experiments. *J. Mod. Opt.* **56**, 1780-1784.
Ortyn, W. E., Piloco, L. R., and Hayenga, J. W. (2000). Cytological system illumination integrity checking apparatus and method. *US Patent* 6011861.
Osvay, K. , Kovács, A. P., Kurdi, K., Heiner, Z., Divall, M., Klebniczki, and Ferincz, I. E. (2005). Measurement of non-compensated angular dispersion and the subsequent temporal lengthening of femtosecond pulses in a CPA laser. *Opt. Commun.* **248**, 201-209.
Nova (2007). *Newton's Dark Secrets*. PBS, Boston.
Nova (2011). *Secrets Beneath the Ice*. PBS, Boston.
Ringwood, A. E. (1978). *Safe Disposal of High Level Nuclear Reactor Waste: a New Strategy*. Australian National University, Canberra.
Popov, S. Y (2007). Private communication.
Primmerman, C. A., Murphy, D. V., Page, D. A., Zollars, B. G., and Barclay, H. T. (1991). Compensation of atmospheric optical distortion using a synthetic beacon. *Nature* **353**, 141-143.
Pryor, A. W. (1997). Personal memories of two advanced enrichment projects at Lucas Heights in the years 1972-1980. *The Australian & New Zealand Physicist* **33** (3-4), 53-58.
Schäfer F. P. (Ed.) (1990). *Dye Lasers*, 3rd Ed., Springer-Verlag, Berlin.
Schrödinger, E. (1992). *What is Life? with Mind and Matter and Autobiographical Sketches*. Cambridge University, Cambridge [Note: this is a reprint of Schrödinger's early essays published in 1944 and 1958, respectively].

Siegner, U., Achermann, M., and Keller, U. (2001). Spatially resolved femtosecond spectroscopy beyond the diffraction limit. *Meas. Sci. Technol.* **12**, 1847-1857.

Singh, S., Dasgupta, K., Kumar, S., Manohar, K. G., Nair, L. G., and Chattergee, U. K. (1994). High-power high-repetition-rate copper-vapor-pumped dye laser. *Opt. Eng.* **33**, 1894-1904.

Sirat, G. Y., Wilner, K., and Neuhauser, D. (2005). Uniaxial crystal interferometer: principles and forecasted applications to imaging astrometry. *Opt. Express* **13**, 6310-6322.

Sliski, A. P. (1997). CCD X-ray microdensitometer system, *US Patent* 5623139.

Sneddon, J., Thiem, T. L., and Lee, Y-I. (eds.) (1996). *Lasers in Analytical Atomic Spectroscopy*. Wiley-VCH, New York.

Sugiyama, A., Nakayama, T., Masaaki, K., Maruyama, Y., and Arizawa, T. (1996). Characteristics of a pressure-tuned single-mode dye laser oscillator pumped by a copper vapor laser. *Opt. Eng.* **35**, 1093-1097.

Ward, J. C. (2004). *Memoirs of a Theoretical Physicist*. Optics Journal, Rochester.

Ward, P. W. and Brownlee D. (2000). *Rare Earth*, Copernicus, New York.

8.15 Appendix I: understanding lasers†

Dedicated to young laser apprentices: sometime ago, I wrote a tunable lasers tutorial, or tunable laser tutorial, that has been fairly well received. Now I have written a succinct summary of the equations [PDF] needed to design narrow-linewidth tunable lasers and predict their coherence characteristics. Unfortunately these tutorials assume certain background in physics and optics which might not be available to most readers. Here I will attempt to describe what is needed in order to build up such a background. In fact, I'll go further. I will describe what you need to know and master in order to become a *laser physicist*. This will be done from the perspective of the most wondrous of lasers: *the tunable laser*. To appreciate what is needed to understand tunable lasers first it must be realized that tunable lasers integrate the knowledge base of various disciplines in physics and mathematics. Here are the three basic components:

Newtonian mechanics: this is the soul of a physicists. Understanding Newtonian mechanics teaches you how to *think*. It also requires a good knowledge of the tools of geometry and calculus. On the practical side you should gain a basic understanding of mechanical engineering and vacuum physics.

Electromagnetism: Maxwell equations in particular and all the mathematical tools needed to handle them. On the practical side you need to know how to design and build transmission lines and high voltage power supplies.

Quantum mechanics: I prefer the Dirac version of quantum physics. Dirac's quantum physics opens the door to understanding the principle of stimulated emission, Heisenberg's Uncertainty Principle, and the essence of *interference* which is the heart of physical optics. Understanding the Schrödinger equation, and its applications, is necessary and knowledge of the density matrix formalism is also useful.

In addition, you will need a background in various branches of mathematics such as matrix algebra, Fourier theory, Laplace transforms, Legendre polynomials, and differential equations. Once the fundamentals are in place you can familiarize yourself with solid state physics and gain media in the liquid and the gas phase.

Now some observations about *optics*: once you have all the basic tools in your archives you should be able to study and master the principles of geometrical and physical optics. Optics is beautiful and this endeavor should be fun. One aspect not mentioned previously, and essential to build new lasers, is *manual ability*. The alignment of a laser requires excellent eye to hand coordination. Optics demands a delicate and clever touch. You should be able to develop this skill interacting with simple opto-mechanical projects. The earlier the better.

Learning and becoming familiar with the topics just outlined can take years of study even after a first university degree in physics or optics is completed. The truth is: you'll never stop learning. However, after some level of competency is reached and after your peers recognize you as a *laser physicist* you will find that you will be able to contribute beyond the boundaries of laser physics. Indeed, you should be able to design and build optically-based instrumentation for industrial, scientific, medical, and astronomical applications. In other words, you will become an excellent experimental physicist with a sound understanding of useful theory. Besides the pleasure and joy of observing your creativity at work you might even encounter the rare privilege of making a few discoveries of your own... and remember always... *walk as children of light* (Ephesians 5).

Suggested Optics and Physics Reading

Born, M. and Wolf, E. (1999). *Principles of Optics*. Cambridge University, Cambridge.

Dirac, P. A. M. (1947). *The Principles of Quantum Mechanics*. Oxford, London.

Duarte, F. J. (2000). Newton, prisms, and the *opticks* of tunable lasers. *Opt. Photon. News* **11** (5), 24-28.
Duarte, F. J. (2003). *Tunable Laser Optics.* Academic-Elsevier, New York.
Duarte, F. J., and Hillman, L. W. (eds.) (1990). *Dye Laser Principles.* Academic, New York.
Feynman, R. P., Leighton, R. B., and Sands, M. (1965). *The Feynman Lectures on Physics,* Vols. I-III. Addison-Wesley, Reading.
Kittel, C. (1971). *Introduction to Solid State Physics.* John Wiley, New York.
Lorrain, P., and Corson, D. (1970). *Electromagnetic Fields and Waves.* Freeman, San Francisco.
Newton, I. (1704). *Opticks.* Royal Society, London.
Newton, I. (1726). *Principia Mathematica.* Royal Society, London.

Addendum

Here I mention two little known facts that have been largely ignored in the literature:

The first description of interference directly applicable to the interference of narrow-linewidth high-power laser beams was prophetically given by Dirac in 1930. Dirac's description of interference was not always understood, or accepted, and is referred to by some as "Dirac's dictum." Dirac's description of interference was explained, using practical laser terminology, in notes that include:

"... interference can be analyzed via the interaction of probability amplitudes. These probability amplitudes are said to be represented by wave functions. Hence, interference can be described via the multiplication of an addition of complex wave functions, with its corresponding complex conjugate... Dirac writes about *a [monochromatic] beam of light consisting of a large number of photons...* In the case of a large number of indistinguishable photons his words are just fine" (Duarte, 1998).

"The Dirac discussion... begins with reference to *a beam of roughly monochromatic light;* then prior to his dictum on interference, he writes about *a beam of light having a large number of photons...* In present terms this is no different than the description of interference due to the interaction of a high-power narrow-linewidth laser beam with a two-beam interferometer" (Duarte, 2003).

In the previous description a key concept is that a beam of monochromatic light is a beam is *indistinguishable photons* which is equivalent to a beam of narrow-linewidth laser emission (Duarte, 2003).

References

Duarte, F. J. (1998). Interference of two independent sources. *Am. J. Phys.* **66**, 662-663.

Duarte, F. J. (2003). *Tunable Laser Optics*. Elsevier-Academic, New York. Chapter 2.

In 1965 Feynman used Dirac's notation to describe the mechanics of population inversion and applied his development to dye molecules prior to the first reports on tunable dye lasers in 1966. Amazing? This is mentioned in *Tunable Laser Optics*, Chapter 1.

Note on laser safety: accidents are preventable. A laser laboratory must always be neatly organized and very clean. A laser laboratory is not a place for chaos or dust. Further, always follow the safety recommendations applicable to the class of laser at hand. Ignoring these observations could have extremely grave consequences.

†First published, by the author, on 2005/03/12, as "Understanding tunable lasers" in:
http://www.opticsjournal.com/understandingtunablelasers.htm

8.16 Appendix II: understanding lasers in simple words†

This brief introduction is a companion article to tunable lasers tutorial, and understanding tunable lasers. The intention is to provide a very brief explanation *of what a laser is* in simple and clear language. The concept essential to a laser is the *emission of light* which is *spatially coherent* and *spectrally coherent*:

- Spatial coherence means that the emission is contained in the form of a beam of light that diverges little as it propagates. In a well designed laser this beam divergence obeys the limit imposed by quantum mechanics known as the uncertainty principle.

- Spectral coherence means that the color of the emission is pure or extremely pure. The purity of the emission is quantified by the bandwidth or linewidth, that is the spread, of the color. The narrower the bandwidth, or linewidth, the higher the spectral coherence.

The radiation described above is emitted by *a laser*. A laser can be a macroscopic, or classical, device that emits quantum mechanical radiation. A laser uses two basic components, the *gain medium*, and the *laser cavity*:

- Gain medium: this is an atomic, or molecular, medium that can be excited either optically or electrically. Once excited this medium emits light. This medium can be in the gas, liquid, or solid state.

- Laser cavity: the gain medium is confined within a laser cavity that also includes optical components such as mirrors, prisms, and gratings. The simplest laser cavity is comprised by two mirrors. More sophisticated laser cavities include mirrors, prisms, and gratings. Generally these more sophisticated laser cavities are known as *lasers resonators* or *laser oscillators*.

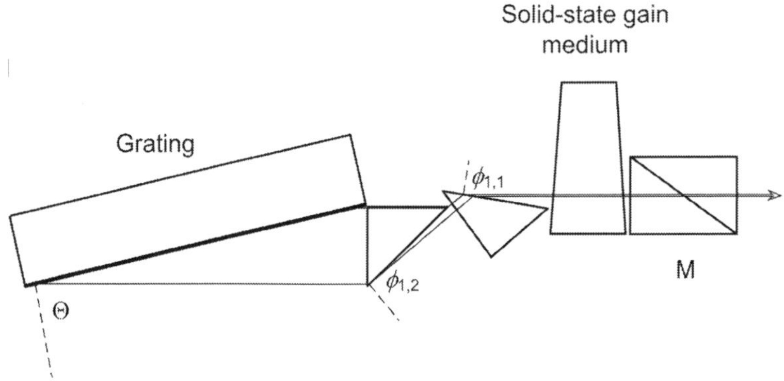

This is an optimized *tunable laser oscillator* yielding extremely pure emission. The schematics show a gain medium, a tuning grating, a multiple-prism beam expander, and an output mirror which is also a polarizer. The emission is tunable in the visible spectrum in the 550-603 nm range (that is, mainly in the yellow-orange portion of the spectrum). The linewidth of this high-power laser oscillator, $\Delta\lambda \approx 0.0004$ nm, is only limited by Heisenberg's uncertainty principle (after Duarte (1999)). Note: as a reference, the visible spectrum covers approximately the region from 380 nm to 770 nm.

With these basic ideas in mind the initial reader should now be more familiar with the concepts described in our laser tutorial. *Tunable Laser Optics* explains and explores further these exciting ideas and provides a simple introduction to the quantum physics behind the laser. Tunable lasers are very widely applied (from astronomy to medicine) and, for those interested, *Tunable Laser Applications* provides an ample authoritative survey also suitable for professionals and university students.

References

Duarte, F. J. (1999). Multiple-prism grating solid-state dye laser oscillator: optimized architecture. *Appl. Opt.* **38**, 6347-6349.

Duarte, F. J. (2003). *Tunable Laser Optics*. Elsevier-Academic, New York.

Duarte, F. J. (2009). *Tunable Laser Applications*. CRC, New York.

†First published, by the author, on 2010/04/16, as "Understanding lasers in simple words" at: *http://www.opticsjournal.com/understandinglasersinsimplewords.htm*

Chapter 9

Observations II

9.1 On coincidences

Newton in his prophetic book *Opticks* (Newton, 1704) described the refraction and dispersion of light. In addition, he hinted at the concept of prismatic expansion and introduced multiple-prism dispersion. Furthermore, he disclosed the reflective telescope which is the basis of the vast majority of telescope designs used today in astronomical observatories around the world. Newton's book also described the basic entity of light, known today as the photon, as a particle. However, *Opticks* corpuscular description of light was abandoned and the accepted description of light became the wave theory of light as advocated by C. Huygens and then A. Fresnel. Hence, Newton's book was ignored for the longest time. Nevertheless, eventually, quantum physics incorporated a dual wave-particle description of light.

One of the creators of quantum mechanics was P. A. M. Dirac. Incidentally, Paul Dirac inherited Newton's chair at Cambridge University. The doctoral supervisor of Dirac was R. H. Fowler who also directed, for a while, M. H. L. Pryce who in turn became the supervisor of J. C. Ward. As mentioned earlier John Ward was the creator of the physics syllabus at Macquarie where I was introduced to quantum mechanics. This introduction was done via the *Feynman Lectures on Physics* (Feynman *et al.*, 1965) which emphasized the Dirac approach to quantum mechanics (Dirac, 1947). Note that quantum mechanics has mainly three avenues of introduction: Heisenberg's, Schrödinger's, and Dirac's.

Around 1980 I was working on my doctoral research. My topic of research was optically pumped molecular lasers in the visible spectrum. To this end, I needed reliable, and very narrow linewidth, tunable lasers. These were not for sale. The best we could do was to build a design introduced by Ted Hänsch then at Stanford (Hänsch, 1972). His configuration used a Galilean telescope to expand the intracavity laser beam and thus achieve the desired linewidth resolution. In this regard, I would spend all morning working on the alignment. Then, came lunch time. In the hot Australian summer, the temperature of the lab would go up a few degrees thus ruining the alignment of the laser due to thermal expansion of the cavity. Thus, on my return from lunch, I had to start again. One day Jim Piper handed me a paper by D. C. Hanna (from Southampton) that disclosed a single-prism beam expander (Hanna *et al.*, 1975). I tried it, and it worked. However, quickly I realized that the main problem with this approach were high reflection losses at the incidence surface of the single prism. I also noticed that the losses were a function of the angle of incidence. Then, it followed that using more than one prism one could achieve the desired intracavity beam expansion with much better efficiencies. I tried prism beam expanders with two and more prisms. They worked like a beauty. I had independently discovered laser intracavity multiple-prism beam expanders (Duarte and Piper, 1980). The multiple-prism near grazing-incidence grating oscillators followed (Duarte and Piper, 1981; 1984a; 1984b).

In 1982, as I was working on my postdoctoral assignment with Brian Orr, at the University of New South Wales, I was also writing my thesis in the evenings. At this stage, it became obvious to me that we did not have a proper dispersion theory to describe multiple-prism beam expanders. Jim Piper said to me something along the lines of: "Look it up in a book." The more I did look the more I realized no one had done it. The only way out was to solve the problem.

In addition, as part of my thesis work, I was writing a paper on new copper-vapor-laser pumped dye lasers. These were multiple-prism oscillators, and would be nice to have the theory, I thought. During the day I would work on my post doc and in the evenings I would work on the multiple-prism theory. Time went by with limited progress. One afternoon, I scheduled a meeting with Jim Piper to discuss the paper in question. Jim reiterated that the important thing was that we had excellent experimental results and then he told me to come back "in about one hour" since he had another meeting. I went to my favorite place, the Macquarie coffee shop and sat in a corner slowly sipping my cappuccino. I had with me a large yellow envelope, from the Australian Institute of Physics. I took my pen and began again.

Then, suddenly, inexplicably, as out of the blue, everything became perfectly clear in my mind. In five minutes I wrote the fundamental steps of what became the generalized multiple-prism grating dispersion theory (Duarte and Piper, 1982). I went back to Jim's office as if walking on air... I had been given "a great gift." With Jim, we went through the equations until he convinced himself. Nevertheless, he told me to check and recheck. Subsequently, we also wrote a pedagogical paper on the subject for the *American Journal of Physics* (Duarte and Piper, 1983). Thus, in 1983, we had a theory to describe linewidth narrowing in multiple-prism grating laser oscillators, and unknowingly, a theory to quantify the dispersion of any prismatic array.

Time went by and in 1987 I extended the theory to higher derivatives to account completely, in a generalized form, for temporal pulse compression in femtosecond lasers (Duarte, 1987). More time went by and in 1997 I wrote a paper that for the first time explained how the Dirac quantum notation besides explaining beautifully and succinctly the physics of interference also leads to a description of diffraction and refraction (Duarte, 1997). Thus Dirac's quantum approach describes the whole of optics in a hierarchal sequence: interference, diffraction, refraction, and reflection (Duarte, 2003, 2004). And of course, refraction is the foundation of prismatic dispersion.

In the late 1990s I came across, for the first time, Newton's *Opticks*. As a seasoned physicist, now I was in awe to learn about Newton's observations on dispersion in a single prism and in multiple-prism arrays. Incredibly, however, he only used words to describe the physics and no equations. I was so impressed by Newton's work that I wrote a little paper, in his honor, entitled *Newton, prisms, and the "opticks" of tunable lasers* (Duarte, 2000). I was also very pleased to realize that our 1982 mathematical description of multiple-prism dispersion was also nicely applicable to Newton's prismatic arrays.

Somehow I had managed to describe the generalized theory of multiple-prism dispersion, which is based on refraction, and the foundations of refraction via Dirac's notation. Was this all just a wonderful coincidence? Now, let us look at the coincidences. First, Dirac, a successor to Newton, played a principal role in providing the physics to describe the photon as having a dual wave-particle nature. Second, John Clive Ward, learned of Dirac's physics via Pryce who did work with Fowler whom also had supervised Dirac. Third, John Ward, who could have easily stayed in the USA, decided to leave for New Zealand. Fourth, one weekend he visited a friend in Sydney. Out of the blue, during that visit he was invited to become foundation professor of physics at Macquarie where he created a physics curriculum based on Dirac's physics. Fifth, I was taken to Australia, by my father, by circumstances totally beyond my control. Sixth, I ended up studying Dirac's physics under John Ward at Macquarie. Seventh, I decided to stay at Macquarie rather than go to Oxford for my doctoral research. Eighth, experimental necessity diverted me to build and improve tunable laser oscillators and this led to the multiple-prism dispersion theory applicable to Newton's arrays (although I did not know it at the time). Ninth, my work on interferometry, stimulated by the imaging needs at Eastman, yielded a multiple-prism *N*-slit interferometer, and allowed me to link Dirac's physics with photon interference, diffraction, *refraction*, reflection, and ultimately with multiple-prism dispersion applicable to Newton's arrays. The circle was now completed (Duarte, 2003). Newton's optics was positioned within the modern framework of quantum physics some 300 years later.

What are the odds of all this just happening by chance? Are all these just random coincidences? At any stage in this consecutive chain of events things could have turned out differently. In fact, the odds of all this turning out as it did are infinitesimally small. And yet… they did. I wonder what would Newton say about this.

9.2 On quantum mechanics

In 1981 I accepted a post doc in laser spectroscopy, with Brian Orr, at New South Wales. At this time I also began to correspond with some US researchers interested in experiments to test Bell's inequalities. When I discussed this with John he said, with his usual frankness: "Don't waste your time." Instead he advised me to work on very sensitive interferometers which, among other applications, would be useful as detectors of gravitational waves. In this regard, like Dirac, he was neither bothered by issues of interpretation nor did he have doubts on the correctness of quantum mechanics. Later when I had to describe a newly invented *N*-slit laser interferometer I did it using Dirac's quantum notation to the dismay of some of my Kodak colleagues. The description is valid for either single-photon illumination or illumination using an ensemble of indistinguishable photons as in the case of narrow-linewidth lasers (Duarte, 1991, 1993). John was pleased with his apprentice. John Ward's thoughts on quantum mechanics are beautifully expressed in his memoirs (Ward, 2004):

"The inner mysteries of quantum mechanics require a willingness to extend one's mental processes into a strange world of phantom possibilities, endlessly branching into more and more abstruse chains of coupled logical networks, endlessly extending themselves forward and even backwards in time."

Note: this section was introduced in Chapter 4 and is reproduced here as background to the following section.

9.3 On the interpretation of quantum mechanics

The interpretation of quantum mechanics has been the subject of many publications and books (see, for example, Bell, 1988; Selleri, 1988; Wallace, 1996). My opinion on this subject has been largely shaped by Ward's teachings and the writings of Dirac and Feynman. Thus, here are the premises:

1. Dirac was not impressed by discussions on the interpretation of quantum mechanics and in one of his last papers he wrote: *"The interpretation of quantum mechanics has been dealt with by many authors, and I do not want to discus it here. I want to deal with more fundamental things"* (Dirac, 1987).

Nevertheless, in a visit to Sydney, in August 1975, Dirac did refer to the interpretation of quantum mechanics and did refer to the Bohr-Einstein controversy. In this regard he went on to say that according to standard atomic physics "Bohr was right." However, he left the possibility open for some kind of improved quantum mechanics of the future. But he warned that a return to determinism would be "at the expense of some other basic idea" (Dirac, 1975).

2. Feynman accepted quantum mechanics as it is. In this regard he wrote: *"the uncertainty principle protects quantum mechanics... quantum mechanics maintains its perilous but still correct existence"* (Feynman *et al.*, 1965).

3. The first correct calculation describing the quantum polarization correlation of two quanta propagating in opposite directions was disclosed in two publications that are almost unknown to the physics community at large (Pryce and Ward, 1947; Ward, 1949). As already mentioned, like Dirac, Ward was not bothered by issues of interpretation.

4. More recently Willis Lamb assigned the interpretational problems of quantum mechanics to "historical misunderstandings" (Lamb, 2001).

A very different perspective is provided in one of the most cited papers in physics in which it is argued, with impeccable and sublime logic, that quantum mechanics is an incomplete theory (Einstein *et al.*, 1935). This paper gave rise to what became known as the Einstein-Podolsky-Rosen (EPR) paradox. Criticisms to quantum mechanics have been numerous and have been generated by some formidable and clever advocates. As recently as 1990, J. S. Bell wrote: "It seems to me that it is among the most sure-footed of quantum physicist, those who have it *in their bones*, that one finds the greatest impatience with the idea that the 'foundations of quantum mechanics' might need some attention" (Bell, 1990).

The story that everyone knows, in an extremely abbreviated version, goes along the following lines: following publication of the EPR argument, Bohm and Aharonov (1957) proposed an experiment to test the EPR hypothesis. This experiment did involve the measurement of polarizations of photons propagating in opposite directions. J. S. Bell then wrote a paper showing that the quantum mechanical treatment of such set up led to results incompatible with local theories (Bell, 1964). Bell's result gave rise to what became known as the celebrated *Bell's inequalities*. Numerous experiments designed to test Bell's inequalities consistently support the quantum results (see, for example, Rowe *et al.*, 2001). Note that these experiments were performed using the polarization of counter propagating photons.

Now, we turn our attention to the work or John Ward and Maurice Pryce. In their extraordinarily succinct *Nature* paper (less than 2/3 of a page) they describe an experimental set up directly applicable to modern EPR type experiments (Pryce and Ward, 1947; Ward, 1949;

Dalitz and Duarte, 2000). In this configuration, positron annihilation yields the emission of two entangled photons propagating in opposite directions. Pryce and Ward (1947) correctly predicted the quantum correlation of the polarizations. Soon, after the Pryce and Ward 1947 paper, measurements by Bleuler and Bradt (1948) and Wu and Shaknov (1950) provided experimental evidence in support of Ward's quantum prediction.

Why does this work remain virtually unknown to the physics community at large? I can think of three possible reasons. First, Pryce and Ward (1947) only disclosed the main result and did not include its derivation. This was described in detail in John's thesis (Ward, 1949) and it was also independently reproduced by Dalitz (1947) and Snyder *et al.* (1948). Second, John Ward never promoted his own work. Third, their *Nature* paper did not refer to the EPR argument and did not make this connection explicit.

An important pause here: Bohm and Aharonov (1957) did refer to Snyder *et al.* (1948) and Wu and Shaknov (1950). Both of these papers, in turn, *did refer* to Pryce and Ward (1947). Furthermore, Bohm and Aharonov (1960) defended the interpretation of the Wu and Shakov experiment "as a test of the paradox of Einstein, Podolsky, and Rosen." If the Wu and Shakov experiment does apply to the EPR paradox so does the theory of Pryce and Ward. In this regard, "Already in 1948, observations of the two photons… agreed with quantum mechanics" Dalitz and Duarte (2000).

The issue of the EPR paradox has endured and came up time and time again when talking to students and colleagues. Thus, around 2001, I wrote a short piece that considers this subject. To do this I went directly back to the paper written by Einstein *et al.* (1935). My argument is the argument of a physicist based on the reality of experimental measurements. This is justified since, in my opinion, physics is essentially an experimental science. In other words, physics without experimental physics and measurements, at its core, does not equal physics.

The argument goes as follows: a recent paper, Lamb (2001) includes several paradoxes and approaches relevant to quantum mechanics in a section entitled "historical misunderstandings."Among the paradoxes mentioned is the Einstein-Podolsky-Rosen paradox. Although much has been written on the EPR paradox there has been a remarkable absence of classical arguments explaining why the EPR paper, albeit a sublime logical argument, represents a misunderstanding as stated by Lamb.

It has been noted, for some time, that the EPR argument is inconsistent with the uncertainty principle. However, authors who have made this point have either not been explicit on why the EPR argument is at odds with the uncertainty principle (Wallace, 1996) or have not stated why the EPR assumptions lead to this inconsistency (Russell, 1985). The aim of this note is to provide a succinct and transparent explanation.

Central to the EPR paper is the issue of incompleteness in quantum mechanics. In this regard, it should be indicated that the first part of the EPR paper (Einstein *et al.*, 1935) states that *"when the momentum of a particle is known, its coordinate has no physical reality."* This would be true only, *and only if,* the momentum p is exactly known. However, in practice we can only determine $p \pm \Delta p$. An uncertainty in p of $\Delta p = 0$ is physically impossible. Uncertainty in measurements is a well established phenomenon in experimental physics since Newton's *Principia* (Newton, 1726).

With a non-zero value for Δp the coordinate can be determined with some degree of certainty according to $\Delta x \approx h/\Delta p$. In other words, Δx has a definite and finite experimental value. This is how Feynman describes the uncertainty principle (Feynman *et al.*, 1965), in itself a well-known experimental reality. The EPR paper, although a master piece of impeccable logical reasoning, invokes an idealistic exact determination of momentum (that is, $\Delta p = 0$) which leads to an *"all value"* spread of Δx. This unphysical *"all value"* spread in Δx results in the early premise of *"no physical reality"* and subsequently in the claim of incompleteness.

At this stage it is appropriate to mention that Bohr (1935) in his reply to EPR was the first to use the uncertainty principle in his argument, nevertheless, he did not provide a clear and direct response to the EPR criticism based on the fact that it is physically impossible to determine p exactly. The note ends here.

Once the argument is made that there is no paradox, using a basic experimental perspective, then the opinions of Dirac, Feynman, Lamb, and Ward come into context. Albeit I did follow the arguments on the interpretation of quantum mechanics, throughout the 1980s and early 1990s, I was busy at work applying Dirac's notation to the description of N-slit interference and obtaining remarkable success in the theoretical confirmation, and prediction, of beautiful experimental results (Duarte, 1991, 1993, 2003).

As a footnote to this discussion I will bring into focus a letter I wrote in defense of the Dirac's description of interference (Dirac, 1947). In that letter (Duarte, 1998) it is explained that Dirac's profound description of interference could even be extended to the experimental situation encountered when an intense narrow-linewidth laser interacts with an N-slit, or multiple-beam, interferometer. That is, Dirac outlined the principles of laser optics back in 1947, in other words: "Dirac wrote the first chapter in laser optics" (Duarte, 2003) and thus he should be considered *the father of quantum optics*.

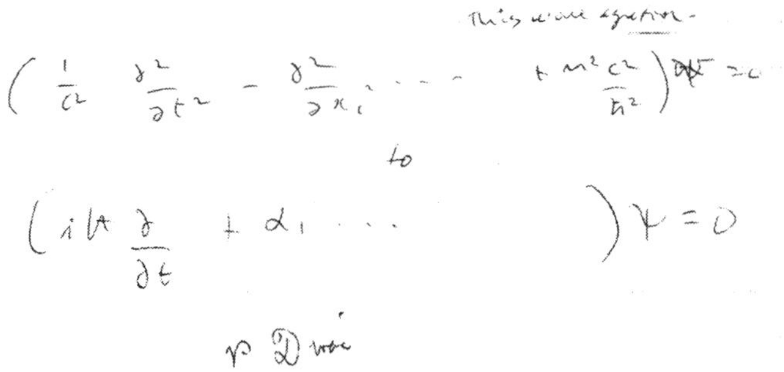

Notes recorded during Dirac's visit to Australia in 1975. The last line shows the only autograph ever collected by the author.

9.4 On reality

Much has been written about reality recently. As an experimental physicist I define reality simply as a set of measurements. Each measurement is a glimpse into reality and the average of a series of measurements, plus or minus its error, an expression of reality.

At a human personal level, reality certainly assumes a much richer texture and might even include experiences outside the four measurable dimensions of space and time. Those altered reality experiences might be the result of chemical stimulation of the brain. Here, I have a confession to make: in my entire life, except for a few cups of good wine, I have never consumed any chemical substance, or drug, to alter my perception of reality. Indeed, I believe that it is a privilege to be able to detect and sense our surroundings using our own unaltered measuring instruments which include our eyes, our ears, our skin, and our sense of smell. Those are the instruments that allow us to gauge reality. The information received is then processed in our brain which decides what to do. However, as aviators know, sometimes the set of sensory mechanisms fail and we must rely on man made instruments to gauge reality and stay alive.

As good as our senses are, they are limited. We see only visible wavelengths which represent a very small portion of the electromagnetic spectrum and we hear only a narrow range of frequencies. To see and hear beyond our sensory limits we have developed an array of detectors and instruments which enable us to broaden our measurable reality.

An alternative way to venture outside the measurable four dimensions of space and time is to experience love, compassion, or to enter the spiritual domain. In my life I have had a few unusual experiences which could be described as "spiritual" but find it very difficult, almost impossible, to explain. In this regard, from a physicists perspective, the best that I have found are Schrödinger's words: "We know when God is experienced this is an event as real as an immediate sense of perception… " (Schrödinger, 1992). This is all I can say about the spiritual side of things.

In the search for alternative interpretations of quantum mechanics, which as explained in the previous section I find unnecessary, authors have ventured into various forms interpretations which include a "many universes interpretation." This concept was first introduced by Everett (1957) and has been more recently articulated by several authors including Deutsch (1997).

First, as a physicist one must always be open to possibilities and accept the doubts brought about by our limited perception. However, one must also ask for the experimental evidence. So far, there is no published experimental evidence that there is more than one universe. In his book Deutsch (1997) uses single photon interference to introduce a conceptual justification of a parallel universe. However, his description of single photon interference is unnecessary given the Dirac description of interference (Dirac, 1947) and recent explanations of this description (Duarte, 1998). In essence, single-photon interference can be described considering the large coherence length of the single photon that allows it to illuminate two or many slits simultaneously. Then, it is the probability amplitudes, interacting according to the mechanics of Dirac that gives rise to a measurable interferometric distribution (Duarte, 1993, 2003, 2004). The equation thus obtained is applicable either to a single photon or to a population of indistinguishable photons. Simultaneous illumination of two or *N*-slits can be easily arranged experimentally. In summary, it is not necessary to introduce the notion of a parallel universe to explain single-photon interference. The mathematical interaction of probability amplitudes corresponds to an interferometric interaction of "wave functions of ordinary wave optics" (Dirac, 1947).

9.5 On the big question I

What did the pillars of modern civilization think about a universal intelligence? What Maxwell had to say? What was the opinion of the creators of quantum mechanics? What was the attitude of the physicists that influenced my own path? Did they care? Were they neutral? Were they atheists?

As mentioned previously modern technology rests mainly on the foundations of electromagnetism and quantum mechanics. Thus, we shall briefly examine the known opinions on this subject of the father of modern physics, *Isaac Newton,* the creator of Maxwell Equations, *James Clerk Maxwell,* and the creators of quantum mechanics: *Paul Dirac, Werner Heisenberg, Max Planck,* and *Erwin Schrödinger.*

I also examine the opinions on the subject of those physicists that have had a direct and indirect influence on my own education: *John Ward,* directly, and *Richard Feynman,* via his textbooks. This is certainly an extremely selective group of physicists and this selection is certainly biased by my opinion as an experimentalist and my own development as a physicist.

In the following description the present tense will be used simply because these physicists produced immortal physics and as such, for this discussion, their opinions are considered timeless.

Paul Dirac: in 1927, young Dirac said as part of a much longer statement: "I don't know why we are talking about religion... The very idea of God is a product of human imagination" (Heisenberg, 1971). Besides this very explicit denunciation of religion, and hence divinity, later in his career he said very little about philosophy or philosophical matters. However, Kragh (1990) indicates that Dirac adheres to the observability doctrine. According to this doctrine physical theories should build on quantities that can be observed. We also know that Dirac was a principal proponent of *mathematical beauty*. In his book Kragh links the principle of aesthetics in mathematics with the concept of a great mathematical architect. In particular, Kragh (1990) cites Dirac as saying in 1963: "God is a mathematician of a very high order, and he used very advanced mathematics in constructing the universe." Kragh editorializes that, coming from Dirac, the word "God" should be interpreted as "nature."

In his 1975 visit to Sydney, Australia, Dirac did pronounce a few statements that, as a young physics student, I faithfully recorded in my notebook. First, on the issue of mathematical beauty he cited W. R. Hamilton and Hamilton's continued use of the mathematical style introduced by Newton. He praised the remarkable mathematical insight of Hamilton and highlighted the importance of the Hamiltonian transformation in quantum mechanics (Dirac, 1975).

Richard Feynman: in one if his latter books, Feynman wrote that he believed that a majority of scientists do not believe in God. From his discussion, it can also be established that he himself did not believe in God. However, in the same writings with his usual eloquence and honesty he goes on to expose his opinion: "... science cannot disprove the existence of God. I absolutely agree. I also believe that a belief in science and religion is consistent" (Feynman, 1998). His exposition continues: "Western civilization...stands by two great heritages... One is the scientific spirit of adventure... The other ... is Christian ethics..." (Feynman, 1998). Further, Feynman emphasizes the central importance of *doubt* in the improvement of physics.

Werner Heisenberg: he was a practicing Lutheran . Further, he thought that religion provided the foundations for ethics (Seeger, 1985). When Pauli asked him if he believed in God his reply is quotes as: "Can you, or anyone else, reach the central order of things... as directly as you can reach the soul of another human being?... If you would put the question like that, the answer is yes." (Seeger, 1985). Heisenberg himself discussed the issue of divinity directly in his book *Physics and Beyond* (Heisenberg, 1971).

James Clerk Maxwell: a devout Christian. However, he did criticize "the misuse of partial scientific knowledge to interpret scripture" (Hutchinson, 2006). Besides his monumental contribution to electromagnetism, Maxwell was also interested in astronomy, atoms, and molecules. In a paper published in 1873, Maxwell reports on his astronomical observations: "We are thus assured that molecules of the same nature as those of our hydrogen exist in those distant regions..." (Maxwell, 1873) and he continues by expressing his belief that time has left unchanged the configuration of molecules: "None of the processes of Nature, since the time since Nature began, have produced the slightest difference in the property of any molecule." (Maxwell, 1873). Here we should indicate that he was referring to molecules such as H_2, O_2, etc. Maxwell died in pain but never abandoned his faith (Hutchinson, 2006).

Isaac Newton: a man immersed in themes of divinity, here are some of his thoughts: "... God in the Beginning form'd Matter in solid... impenetrable, movable Particles... even so hard, as never to wear or break in pieces; no ordinary power being able to divide what God himself made one in the first Creation" (Newton, 1704). He makes further statements on this topic, notably: "God is able to create Particles of... several Sizes... and perhaps of different Densities and Forces..." (Newton, 1704). Considering Newton's mathematical genius, it is not hard to imagine that, given some experimental evidence, he could have easily discovered particle physics back in 1704.

Max Planck: "...Planck considers religion and science compatible because... they refer to quite different facets of reality" (Heisenberg, 1971).

Erwin Schrödinger: he did follow Catholicism and he was also involved with the teachings of Vedanta (Moore, 1989). As such, he did publish some provocative thoughts on the issue of divinity and did refer to a "supreme mind." In this regard he wrote: "We know when God is experienced this is an event as real as an immediate sense of perception... God is spirit" (Schrödinger, 1992). However, from my own perspective, what I find most enlightening in his general writings is the following sentence: "A mathematical truth is timeless, it does not come into being when we discover it..." (Schrödinger, 1992).

John Ward: in all my conversations with him he said almost nothing about matters of divinity. Certainly, he did not participate in organized religion. Also, his sister told me that he requested no religious service when facing his end. However, back at Macquarie, I do remember that in various occasions he did praise, with an oblique sense of humor, King Henry the VIII for "having created his own church." Beyond this peripheral point, in John's perspective, the fight between good and evil was very real.

In one of the last conversations I had with him, he did refer to entropy as the "ultimate evil." In previous conversations he would also repeatedly mention his "guardian angel" as documented in his writings (Ward, 2004).

9.6 On the big question II

Next to these giants my opinion's weight is infinitesimal. However, I will allow myself to make some observations. I should begin by stating that I consider the disbelief on the existence of a deity quite acceptable. Indeed the right to disbelieve, or believe, is central to the concept of a free society.

At this stage, I should consider the suffering of Galileo Galilei, at the hands of the inquisition. As a physicist, and a nonconformist, I sincerely deplore Galileo's trial and imprisonment. However, that was not an issue of Galileo versus Christianity. That was an issue of organized religion exercising political power. I have little doubt that, had Galileo been an Englishman, living in England, his fate would have been quite different.

Parenthetically, on this issue, I should clarify that I don't participate in organized religion, although I do accept my Christian heritage. Also, I find the excesses in any type of dogma, simply abhorrent. The abuse and misuse of power, religious or political, should be condemned by all those who love freedom and seek truth.

It should be noted that, for a long time, I have held the belief that issues of faith and science should be kept in two different planes. In this perspective, science deals with the measurable, and divinity with the spiritual, a view perfectly compatible with the notion attributed to Planck (Heisenberg, 1971). To me, this was always an intuitive notion, to which I gave no further thought since my days at Macquarie.

Modern science, as we know it, has its origins in Newton's era. Extending Feynman's observation, and expanding on Heisenberg's position on ethics, I have to say that Christianity has been central in making available an ethical framework beneficial to the practice of modern science. Within this framework scientists have been trained to report findings truthfully and honestly. Without this ethical framework science, as we know it, would not have been possible.

A further observation has to do with the gift of discovery. In this regard, in a few occasions I have been privileged to observe discovery with my own eyes. I do clearly remember the sense of awe… when I aligned into oscillation my first tunable laser design. I have experienced a similar sensation of elevated consciousness while discovering new equations. As Schrödinger (1992) said, it is… *"like a great gift."*

Based on Newton's physics, Maxwell equations, and quantum mechanics, a plethora of wonderful technologies have emerged: the transistor, and hence the modern computer, the internet, the digital phone, the laser, the tunable laser, nanotechnology, very large telescopes, NMR, and then MRI. These technologies have had an enormous impact in astronomy, biology, communications, chemistry, medicine, and society in general. Modern medicine is one of the avenues from which the public has received perhaps some of the most recognizable benefits. Today, medical practitioners can observe the body, its inner workings, and the brain, in ways unimaginable not long ago. This physics derived technology has enabled great advances and will continue to do so. In a different area of endeavor, astronomers, thanks to Newton and quantum mechanics, can pierce deeper into the heavens more than ever before.

Back to the early Dirac statement: "The very idea of God is a product of human imagination" (Heisenberg, 1971). Here it should be pointed out, however gently, that a great deal of very good physics has been discovered thanks to "human imagination." Quantum mechanics itself is the child of human imagination and a great triumph of human imagination. All the wonderful technology we have is a product of human imagination. In fact, Dirac himself had a most wonderful human imagination and his imagination enabled him to discover the elegant physics he gave us.

Indeed, science and human imagination, might eventually create totally artificial forms of life thus turning scientists into *nano gods*. In this regard, the works of the great physicists have made possible the emergence of nanogods. Then, if man is capable of turning himself into a creator, is it not possible that, following the Big Bang, mankind itself was pre-programmed as a subcomponent of the universe?

As mentioned previously, Feynman exalts the value of *doubt* and considered it "essential to progress" (Feynman, 1998). In my view, the great lesson from Feynman's perspective is that we have to keep an open mind.

In a recent book, Hawking wrote: "Because there is a law such as gravity, the universe can and will create itself from nothing" (Hawking and Mlodinow, 2010). That is fine, however, the laws of gravity are mathematical laws, and the question then becomes, who wrote the laws of gravity? Did these laws created themselves? And if so, who programmed this self creation?

Now, regardless of the origin of the universe: if biological life is just the result of probabilistic laws, are those the laws of quantum mechanics? Are those the laws of Dirac's divine mathematician?

And if there is a grand mathematician... will we ever understand how he came to be? Perhaps behind these questions there is a yet undiscovered, and perhaps unimaginable truth. And the human journey towards the ultimate revelation will continue to be a fascinating quest.

As a physicist, besieged by doubt and at the same time motivated by doubt, I think that Planck's and Feynman's duality is perfectly acceptable. However, at the same time, I find Schrödinger's observation *"A mathematical truth is timeless, it does not come into being when we discover it..."* (Schrödinger, 1992) rather humbling and profoundly inspiring.

9.6 On the big question III

A third aspect of this theme has to do with the question: why are we here?

Is our existence simply the result of some random event? Is our function simply to be born, reproduce, and die? Is the future of life simply death? In other words, are we humans just an insignificant, and irrelevant, byproduct of the universe?

Or… do humans represent a deeper purpose? Is our function to explore, discover, and understand the universe? And where does the urge to explore and discover come from? Certainly, we don't know the answer to these questions. But, if our innate urge to explore is part of the human program, why is it so? To admire the awesome beauty of the universe? To discover the sublime mathematical truths mentioned by Schrödinger? Who wrote the program? Will science ever provide answers to these questions?

References

Bell, J. S. (1964). On the Einstein-Podolsky-Rosen paradox. *Physics* **1**, 195-200.

Bell, J. S. (1964). *Speakable and Unspeakable in Quantum Mechanics.* Cambridge University, Cambridge.

Bell, J. S. (1990). Against measurement. *Physics World* (8), 33-40.

Bleuler, E., and Bradt, H. L. (1948). Correlation between the states of polarization of the two quanta of annihilation radiation, *Phys. Rev.* **73**, 136.

Bohm, D., and Aharonov, Y. (1957). Discussion of experimental proof for the paradox of Einstein, Rosen, and Podolsky. *Phys. Rev.* **108**, 1070-1076.

Bohm, D., and Aharonov, Y. (1960). Further discussion of possible experimental tests for the paradox of Einstein, Podolsky and Rosen, *Nouvo Cimento* **17**, 964-976.

Bohr, N. (1985). Can quantum mechanical description of physical reality be considered complete? *Phys. Rev.* **48**, 696-702.

Dalitz. R. H. (1947). Theoretical physics seminar delivered at the University of Cambridge in October 1947.

Dalitz, R. H., and Duarte, F. J. (2000). John Clive Ward. *Physics Today* **53** (10), 99.

Deutsch, D. (1997). *The Fabric of Reality: The Science of Parallel Universes and its Implications.* Allen Lane, New York.

Dirac, P. A. M. (1947). *The Principles of Quantum Mechanics.* 3rd Ed., Oxford, London.

Dirac, P. A. M. (1975). Lecture delivered at the University of New South Wales, Sydney, Australia.

Dirac, P. A. M. (1987). The inadequacies of quantum field theory. In *Paul Adrien Maurice Dirac* (Kursunoglu, B. N., and Wigner, E. P., eds.). Cambridge, London. Chapter 15.

Duarte, F. J. (1987). Generalized multiple-prism dispersion theory for pulsed compression in ultrafast dye lasers. *Opt. Quantum Electron.* **19**, 223-229 (1987).

Duarte, F. J. (1991). Dispersive dye lasers. In *High Power Dye Lasers* (Duarte, F. J., ed.). Springer-Verlag, Berlin, Chapter 2.

Duarte, F. J. (1997). Interference, diffraction, and refraction, via Dirac's notation. *Am. J. Phys.* **65**, 637-640.

Duarte, F. J. (1998). Answer to question 60: interference of two independent sources. *Am. J. Phys.* **66**, 662-663.

Duarte, F. J. (2000). Newton, prisms, and the *opticks* of tunable lasers. *Opt. Photon. News.* **11** (5), 24-28.

Duarte, F. J. (2003). *Tunable Laser Optics.* Elsevier Academic, New York.

Duarte, F. J. (2004). Comment on "Reflection, refraction, and multislit interference." *Eur. J. Phys.* **25**, L57-L58.

Duarte, F. J. and Piper, J. A. (1980). A double-prism beam expander for pulsed dye lasers. *Opt. Commun.* **35**, 100-104.

Duarte, F. J., and Piper, J. A. (1981). A prism preexpanded grazing incidence pulsed dye laser. *Appl. Opt.* **20**, 2113-2116.

Duarte, F. J. and Piper, J. A. (1982). Dispersion theory of multiple-prism beam expander for pulsed dye laser. *Opt. Commun.* **43**, 303-307.

Duarte, F. J. and Piper, J. A. (1983). Generalized prism dispersion theory. *Am. J. Phys.* **51**, 1132-1134.

Duarte, F. J., and Piper, J. A. (1984a). Narrow linewidth high prf copper laser-pumped dye-laser oscillators. *Appl. Opt.* **23**, 1391-1394 (1984).

Duarte, F. J., and Piper, J. A. (1984b). Multi-pass dispersion theory of prismatic pulsed dye lasers. *Optica Acta* **31**, 331-335.

Einstein, A., Podolsky, B., and Rosen N. (1935). Can quantum mechanical description of physical reality be considered complete? *Phys. Rev.* **47**, 777-780.

Everett, H. (1957). "Relative state" formulation of quantum mechanics. *Rev. Mod. Phys.* 29, 454-462.

Feynman, R. P. (1998). *The Meaning of it All*. Addison-Wesley, Reading.

Feynman, R. P., Leighton, R. B., and Sands, M. (1965). *The Feynman Lectures on Physics*, Vol. III. Addison-Wesley, Reading.

Hanna, D. C., Karkkainen, P. A., and Wyatt, R. (1975). A simple beam expander for frequency narrowing of dye lasers, *Opt. Quantum Electron.* **7**, 115-119.

Hänsch, T. W. (1972). Repetitively pulsed tunable dye laser for high resolution spectroscopy, *Appl. Opt.* **11**, 895-898.

Hawking, S., and Mlodinow L. (2010). *The Grand Design*, Random House, New York.

Heisenberg, W (1971). *Physics and Beyond*, Harper and Rowe.

Hutchinson, I. (2006). *James Clerk Maxwell and the Christian Proposition*. http://silas.psfc.mit.edu/maxwell

Kragh, H. S. (1990), *Dirac: A Scientific Biography*. Cambridge, London.

Lamb, W. E. (2001). Super classical quantum mechanics: the best interpretation of nonrelativistic quantum mechanics. *Am. J. Phys.* **69**, 413-421.

Maxwell, J. C. (1873). Molecules. *Nature* **8**, 437-442.

Moore, W. (1989). *Schrödinger Life and Thought.* Cambridge University, London.

Newton, I. (1704). *Opticks,* Royal Society, London.

Newton, I (1726). *Principia Mathematica.* Royal Society, London.

Pryce, M. H. L., and Ward, J. C. (1947). Angular correlation effects with annihilation radiation. *Nature* **160**, 435.

Rowe, J. S., Kielpinski, D., Meyer, V., Sackett, C. A., Itano, W. M., Monroe, C., and Wineland, J. (2001). Experimental violation of a Bell's inequality with efficient Detection. *Nature* **409**, 791-794.

Russell, B. R. (1985). Preparation-limited predictions in Einstein-Podolsky-Rosen experiments. *Found. Phys.* **15**, 861-869.

Salleri, F. (Ed.) (1988). *Quantum Mechanics Versus Local Realism.* Plenum, New York.

Schrödinger, E. (1992). *What is Life? with Mind and Matter and Autobiographical Sketches,* Cambridge University, Cambridge [Note: this is a reprint of Schrödinger's early essays published in 1944 and 1958, respectively].

Seeger, R. J. (1985). Heisenberg, thoughtful Christian. *J. Am. Sci. Affiliation* **37**, 231-232.

Snyder H. S., Pasternack, S., and Hornbostel, J. (1948). Angular correlations of scattered annihilation radiation. *Phys. Rev.* **73**, 440-447.

Wallace, P. R. (1996). *Paradox Lost.* Springer, Berlin.

Ward, J. C. (1949). *Some Properties of the Elementary Particles,* Oxford University, D. Phil. Thesis.

Ward, J. C. (2004). *Memoirs of a Theoretical Physicist.* Optics Journal, Rochester.

Wu, C. S., and Shaknov, I. (1950). The angular correlation of scattered annihilation radiation. *Phys. Rev.* **77**, 136.

Index

16707511R00128

Made in the USA
Lexington, KY
09 August 2012